An Illustrated Review of the

SKELETAL AND MUSCULAR SYSTEMS

Glenn F. Bastian

 Addison-Wesley Educational Publishers, Inc.

Executive Editor: Bonnie Roesch
Cover Designer: Kay Petronio
Production Manager: Bob Cooper
Printer and Binder: Malloy Lithographing, Inc.
Cover Printer: The Lehigh Press, Inc.

AN ILLUSTRATED REVIEW OF THE SKELETAL AND MUSCULAR SYSTEMS

by Glenn F. Bastian

Library of Congress Cataloging-in-Publication Data

Bastian, Glenn F.
        An illustrated review of the skeletal and muscular systems / Glenn
    F. Bastian.
            p.   cm.
        Includes bibliographical references.
        ISBN: 0-06-501704-8
        1. Musculoskeletal system—Anatomy—Outlines, syllabi, etc.
    2. Musculoskeletal system—Anatomy—Atlases. I. Title. II. Title:
    Illustrated review of the skeletal and muscular systems.
        [DNLM: 1. Musculoskeletal System—anatomy & histology—atlases.
    2. Musculoskeletal System—physiology—outlines. WE 17 B3258i 1993]
    QM100.B38 1993
    611'.7—dc20
    DNLM/DLC
    for Library of Congress   93–25475
                                                                              CIP

97 98 99 00 9 8 7 6

*In Memory
of
My Parents*

CONTENTS

All illustrations used in Part I are unlabeled for self-testing.

LIST OF TOPICS & ILLUSTRATIONS

Text: One page of text is devoted to each of the following topics. *Illustrations are listed in italics.*

PREFACE

An Illustrated Review of Anatomy and Physiology is a series of ten books written to help students effectively review the structure and function of the human body. Each book in the series is devoted to a different body system.

My objective in writing these books is to make very complex subjects accessible and unthreatening by presenting material in manageable size bits (one topic per page) with clear, simple illustrations to assist the many students who are primarily visual learners. Designed to supplement established texts, they may be used as a student aid to jog the memory, to quickly recall the essentials of each major topic, and to practice naming structures in preparation for exams.

INNOVATIVE FEATURES OF THE BOOK

(1) Each major topic is confined to one page of text.

A unique feature of this book is that each topic is confined to one page and the material is presented in outline form with the key terms in boldface or italic typeface. This makes it easy to quickly scan the major points of any given topic. The student can easily get an overview of the topic and then zero in on a particular point that needs clarification.

(2) Each page of text has an illustration on the facing page.

Because each page of text has its illustration on the facing page, there is no need to flip through the book looking for the illustration that is referred to in the text ("see Figure X on page xx"). The purpose of the illustration is to clarify a central idea discussed in the text. The images are simple and clear, the lines are bold, and the labels are in a large type. Each illustration deals with a well-defined concept, allowing for a more focused study.

PHYSIOLOGY TOPICS (1 text page : 1 illustration)
Each main topic in physiology is limited to one page of text with one supporting illustration on the facing page.

ANATOMY TOPICS (1 text page : several illustrations)

For complex anatomical structures a good illustration is more valuable than words. So, for topics dealing with anatomy, there are often several illustrations for one text topic.

(3) Unlabeled illustrations have been included.

In Part II, all illustrations have been repeated without their labels. This allows a student to test his or her visual knowledge of the basic concepts.

(4) A Pronunciation Guide has been included.

Phonetic spelling of unfamiliar terms is listed in a separate section, unlike other textbooks where it is usually found in the glossary or spread throughout the text. The student may use this guide for pronunciation drill or as a quick review of basic vocabulary.

(5) A glossary has been included.

Most textbooks have glossaries that include terms for all of the systems of the body. It is convenient to have all of the key terms for one system in a single glossary.

ACKNOWLEDGMENTS

I would like to thank the reviewers of the manuscript for this book who carefully critiqued the text and illustrations for their effectiveness: William Kleinelp, Middlesex County College, Jean Helgeson, Collin County Community College, and Robert Smith, University of Missouri, St. Louis and St. Louis Community College, Forest Park. Their help and advice is greatly appreciated. I am greatly indebted to my editor Bonnie Roesch for her willingness to try a new idea, and for her support throughout this project. I invite students and instructors to send any comments and suggestions for enhancements or changes to this book to me, in care of HarperCollins, so that future editions can continue to meet your needs.

Glenn Bastian

1 Skeletal System

SKELETAL SYSTEM / Overview

DIVISIONS OF THE SKELETAL SYSTEM   *206 bones*

Axial Skeleton   *80 bones*

Skull : 22 bones (facial bones and cranial bones)

Auditory Ossicles : 6 bones (bones of the middle ear : malleus, incus, and stapes)
> The auditory ossicles are three tiny bones that extend across each middle ear. They are attached to the walls of the middle ear by ligaments and muscles and are connected to each other by synovial joints. They transmit sound waves from the eardrum to the internal ear.

Hyoid Bone : 1 bone (in the neck)
> The hyoid bone is a U–shaped bone located in the neck between the mandible and the larynx. It is suspended from the styloid processes of the temporal bones by ligaments and muscles. It supports the tongue and provides attachment for some of its muscles and for muscles of the neck and larynx. It does not articulate with any other bone.

Vertebral Column : 26 separate bones (33 vertebrae; 5 sacral and 4 coccygeal are fused)

Thoracic Cage : 25 bones (24 ribs, 1 sternum)

Appendicular Skeleton   *126 bones*

Pectoral Girdles (shoulder girdles) : 4 bones (2 clavicles, 2 scapulae)

Pelvic Girdle (hip girdle) : 2 bones (coxal bones)

Upper Extremities (upper limbs) : 60 bones
> 6 arm bones : humerus (2), ulna (2), radius (2)
> 16 carpals (wrist bones)
> 10 metacarpals
> 28 phalanges (finger bones)

Lower Extremities (lower limbs) : 60 bones
> 8 leg bones : femur (2), tibia (2), fibula (2), patella (2)
> 14 tarsals (ankle bones)
> 10 metatarsals
> 28 phalanges (toe bones)

TYPES OF BONES

(1) Long : bones of the arms and legs.

(2) Short : bones of the fingers and toes.

(3) Flat : bones of the cranium, face, and scapulae.

(4) Irregular : bones of the wrists and ankles; auditory ossicles; vertebrae.

(5) Sutural (Wormian) : found between sutures of certain cranial bones.

(6) Sesamoid : develop in tendons or ligaments; patellae (kneecaps).

BONE FUNCTIONS

(1) Support : the skeleton provides a framework for all the structures of the body.

(2) Protection : bone protects soft internal organs (cranium protects the brain).

(3) Movement : bones and joints form lever systems acted on by muscles.

(4) Mineral Homeostasis : bone is the major reservoir for calcium.

(5) Blood Cell Production : blood cells and platelets are formed in the red marrow.

(6) Energy Storage : adipose tissue stores fat in the yellow bone marrow.

SKELETON : Overview

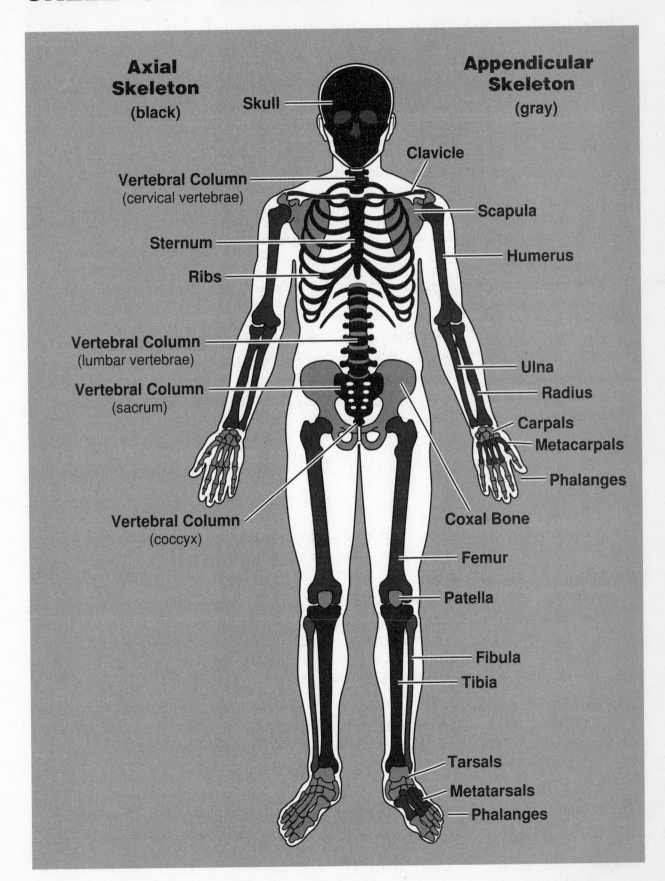

Axial Skeleton (black)

Appendicular Skeleton (gray)

Skull

Vertebral Column (cervical vertebrae)

Clavicle

Scapula

Humerus

Sternum

Ribs

Vertebral Column (lumbar vertebrae)

Vertebral Column (sacrum)

Ulna

Radius

Carpals

Metacarpals

Phalanges

Vertebral Column (coccyx)

Coxal Bone

Femur

Patella

Fibula

Tibia

Tarsals

Metatarsals

Phalanges

SKELETAL SYSTEM / Long Bone Structures

A long bone is one that has greater length than width, such as the tibia (the shinbone of the leg).

Articular Cartilage
The articular cartilage is a thin layer of hyaline cartilage that covers the epiphyses (ends) of a long bone. It reduces friction and absorbs shock at freely movable joints.

Compact Bone or Dense Bone
Compact bone forms the bulk of the diaphyses (shafts) of long bones; it also covers the spongy bone in the epiphyses. Its dense structure gives strength and rigidity to bones.

Diaphysis or Shaft
The diaphysis or shaft is the long, main portion of a long bone. It contains a space called the medullary (marrow) cavity which contains fatty yellow marrow.

Endosteum
The endosteum is the membrane that lines the medullary cavity; consists of a single layer of osteoprogenitor cells.

Epiphyses (singular : epiphysis)
The epiphyses are the ends of a long bone.

Epiphyseal Line
The epiphyseal line is a bony structure that replaces the epiphyseal cartilage (epiphyseal plate); when it appears, the bone stops growing.

Medullary Cavity (Marrow Cavity)
The medullary cavity is a space within the diaphysis of long bones; contains yellow marrow (stores fat).

Metaphysis
The metaphysis is the region between the diaphysis and the epiphysis where calcified matrix is replaced by bone. The bone grows in length as a result of this activity.

Nutrient Foramen
A nutrient foramen is a hole in the diaphysis. Nutrient arteries pass through nutrient foramina in the compact bone; they supply the spongy bone and bone marrow in the diaphysis. Periosteal arteries enter the diaphysis at many locations and supply blood to the compact bone. Epiphyseal arteries and metaphyseal arteries that arise from blood vessels in the associated joints supply the epiphyses of long bones.

Periosteum
The periosteum is a membrane that covers the outer surface of a bone; it has two layers :
> Fibrous Layer (outer layer) : consists of fibroblasts and collagen fibers.
> Osteogenic Layer (inner layer) : contains osteoprogenitor cells.

Spongy Bone
Spongy bone makes up most of the tissue of the epiphyses of long bones. The epiphyses contain red bone marrow, where blood cell production (hemopoiesis) occurs.

LONG BONE STRUCTURES
Tibia (longitudinal section)

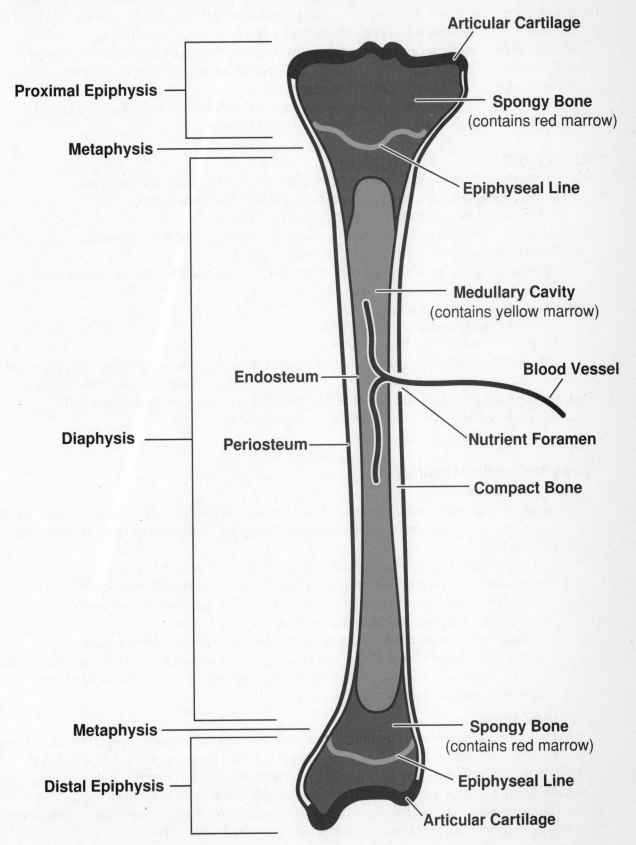

Articular Cartilage

Proximal Epiphysis

Spongy Bone
(contains red marrow)

Metaphysis

Epiphyseal Line

Medullary Cavity
(contains yellow marrow)

Endosteum

Blood Vessel

Diaphysis

Periosteum

Nutrient Foramen

Compact Bone

Metaphysis

Spongy Bone
(contains red marrow)

Distal Epiphysis

Epiphyseal Line

Articular Cartilage

5

SKELETAL SYSTEM / Bone Histology

BONE MATRIX

The matrix is the ground substance and fibers that surround the cells in connective tissues. In bone the ground substance is primarily tricalcium phosphate and the fibers are collagen fibers.

Mineral Salts : mostly tricalcium phosphate (hydroxyapatite); some calcium carbonate, and small amounts of magnesium hydroxide, fluoride, and sulfate are also present; the hardness of bone is due to the crystallization of the inorganic mineral salts.

Fibers : mostly collagen fibers; they form a framework of fibers that make the bone tissues resistant to stretching and tearing, less brittle, and more pliable.

CELL TYPES

In bone tissue the cells are widely separated throughout the matrix. There are four types.

(1) Osteoprogenitor Cells (unspecialized cells derived from mesenchyme)
function : divide by mitosis and develop into osteoblasts.
locations : inner layer of periosteum; endosteum; central and perforating canals.

(2) Osteoblasts (cells derived from osteoprogenitor cells)
function : form bone tissue by secreting matrix (collagen and other organic compounds).
location : surfaces of bone.

(3) Osteocytes (mature bone cells)
function : maintenance of the bone matrix.
location : one osteocyte in each lacuna.
filopodial processes : osteocyte processes that pass through canaliculi (minute channels) and make contact with adjacent osteocytes; provide a pathway for nutrients from the blood.

(4) Osteoclasts (formed by the fusion of 2 to 50 monocytes, a type of white blood cell)
function : resorption of bone (destruction of the matrix).
location : surfaces of bone.

TYPES OF BONE TISSUES

Compact Bone

Structure Osteons (*Haversian systems*) are the basic structural units of compact bone. Each osteon consists of a central canal with its surrounding lamellae (matrix), lacunae, osteocytes, and canaliculi.

Central Canal : a longitudinal canal; contains blood vessels and nerves.
Lamellae : concentric rings of hard, calcified matrix surrounding central canals.
Lacunae : small spaces between the lamellae; each lacuna contains one osteocyte.
Osteocytes : mature bone cells that maintain the bony matrix.
Canaliculi : minute channels that house the filopodial processes of osteocytes.

Location : the diaphyses (shafts) of long bones; it covers spongy bone in the epiphyses (ends). In other types of bones (short, flat, or irregular) it lies over the spongy bone.

Functions : protects, supports, and resists stress.

Spongy Bone

Spongy bone makes up most of the bone tissue of short, flat, and irregularly shaped bones and most of the epiphyses (ends) of long bones. It consists of *lamellae* (layers of matrix) arranged in an irregular latticework of thin plates of bone called *trabeculae*. In the hipbones, ribs, sternum, vertebrae, skull, and the epiphyses of the humerus and femur the spaces between the trabeculae are filled with red bone marrow (the site of blood cell production). *Osteocytes* lie in spaces (*lacunae*), which have minute channels (*canaliculi*) that radiate out into the lamellae.

BONE TISSUE

Epiphysis (end) of a Long Bone
Longitudinal Section

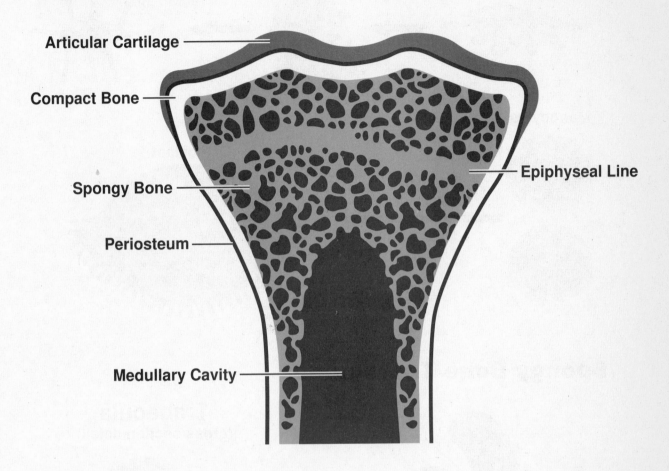

Articular Cartilage

Compact Bone

Spongy Bone

Periosteum

Medullary Cavity

Epiphyseal Line

An Osteon of Compact Bone
Cross Section

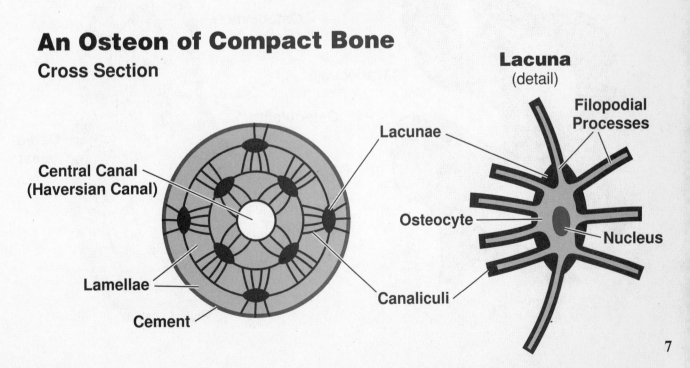

Central Canal
(Haversian Canal)

Lamellae

Cement

Lacunae

Canaliculi

Lacuna
(detail)

Filopodial
Processes

Osteocyte

Nucleus

BONE TISSUE : Cell Types and Spongy Bone

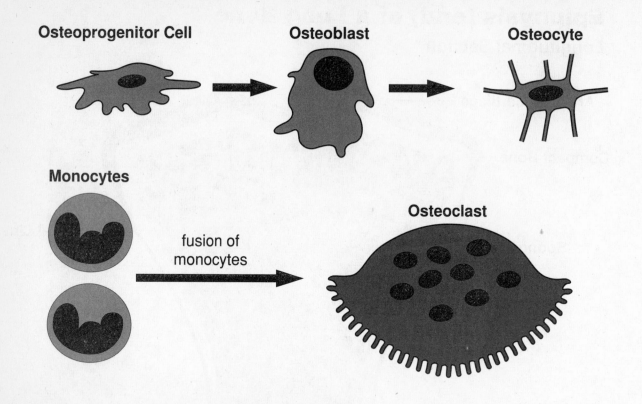

Osteoprogenitor Cell

Osteoblast

Osteocyte

Monocytes

fusion of monocytes

Osteoclast

Spongy Bone Trabeculae

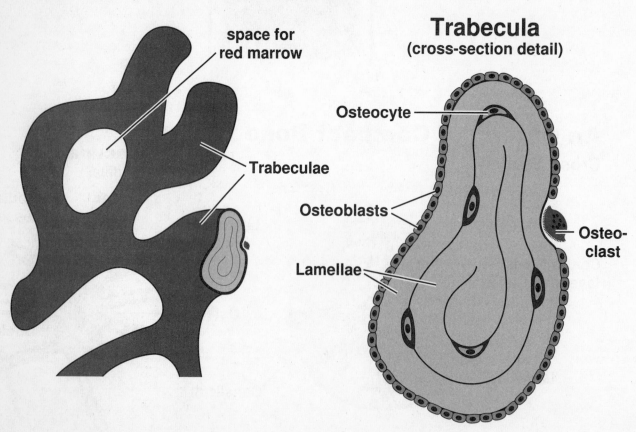

space for red marrow

Trabeculae

Trabecula
(cross-section detail)

Osteocyte

Osteoblasts

Lamellae

Osteo-clast

8

BONE TISSUE : Compact Bone

Cross Section
Diaphysis (Shaft) of a Long bone

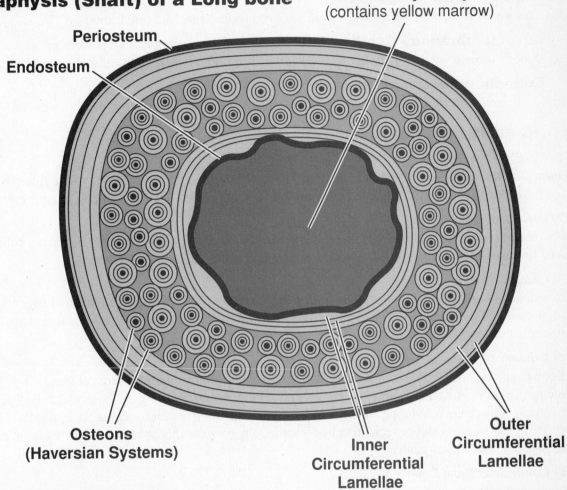

Periosteum

Endosteum

Medullary Cavity
(contains yellow marrow)

Osteons
(Haversian Systems)

Inner
Circumferential
Lamellae

Outer
Circumferential
Lamellae

Longitudinal Section
detail showing osteon structure

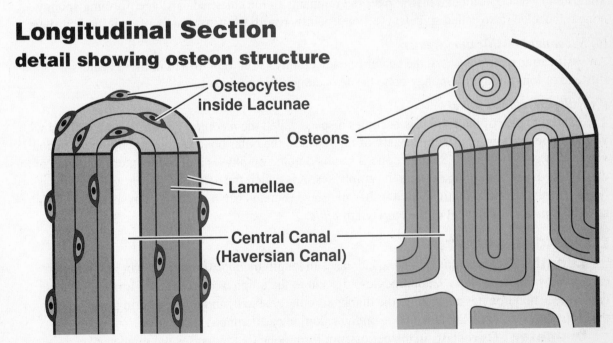

Osteocytes
inside Lacunae

Osteons

Lamellae

Central Canal
(Haversian Canal)

9

SKELETAL SYSTEM / Bone Formation and Growth

TYPES OF BONE FORMATION

During bone formation pre-existing connective tissue is replaced with bone. The process by which bone is formed is called *ossification*. There are two mechanisms for bone formation :

Intramembranous Ossification   (*intra* = within; *membranous* = membrane)
Bone is formed within a membrane; it occurs within fibrous membranes of embryo and adult.

Endochondral Ossification   (*endo* = within; *chondro* = cartilage)
Bone is formed by the replacement of cartilage; it occurs within a cartilaginous model.

LONG BONE DEVELOPMENT

(1) Cartilaginous Model
During embryonic development, mesenchymal cells cluster together in the shape of the future bone. The mesenchymal cells differentiate into chondrocytes that secrete matrix, forming hyaline cartilage.

(2) Periosteal Bone Collar
The membrane surrounding the cartilaginous model is called the perichondrium. It forms a collar of bone tissue (by intramembranous ossification) that surrounds the diaphysis (shaft).

(3) Calcification of Cartilage
Chondrocytes in the midregion of the model increase in size and burst. A change in pH triggers the calcification of the cartilage. The diaphysis is filled with calcified cartilage that contains many large spaces (lacunae) left by the disintegrating chondrocytes.

(4) Primary Ossification Center
Blood capillaries penetrate the calcified cartilage; they carry bone marrow stem cells and osteoprogenitor cells into the spaces. The stem cells proliferate, forming red bone marrow. The osteoprogenitor cells proliferate and differentiate into osteoblasts, which secrete bone matrix. The blood capillaries and osteoprogenitor cells are called the *periosteal bud*; they produce a region called the *primary ossification center*.

(5) Continuous Bone Layer
The osteoblasts deposit bone matrix over the remnants of the calcified cartilage, forming spongy bone trabeculae (also called spicules). Chondroclasts resorb (destroy matrix of) calcified cartilage.

(6) Secondary Ossification Centers
An ossification center arises at the center of each epiphysis (end). Growth in these centers is radial (instead of longitudinal). Spongy bone tissue occupies the epiphyses.

(7) Fully Developed Bone
The central region of the diaphysis becomes a space called the *medullary cavity* which is filled with yellow bone marrow, consisting largely of fat cells. In the outer bone surface spongy bone is converted into compact bone. Spongy bone is retained in the epiphyses and in the interior surface of the diaphysis. Hyaline cartilage covers the epiphyses; it is called the *articular cartilage*. As long as bone growth continues (until about age 20), cartilage remains between the diaphysis and the epiphyses as the major component of the *epiphyseal plate*.

LONG BONE GROWTH

Length   The diaphysis of a bone increases in length by *appositional growth* in the region called the *epiphyseal plate*. Chondrocytes adjacent to the epiphyses proliferate, forming articular cartilage. Chondrocytes adjacent to the diaphysis enlarge, burst, and die, triggering calcification of the matrix. Osteoblasts lay down bone on the calcified cartilaginous trabeculae.

Diameter   Osteoblasts in the periosteum form bone that increases the diameter.

LONG BONE DEVELOPMENT

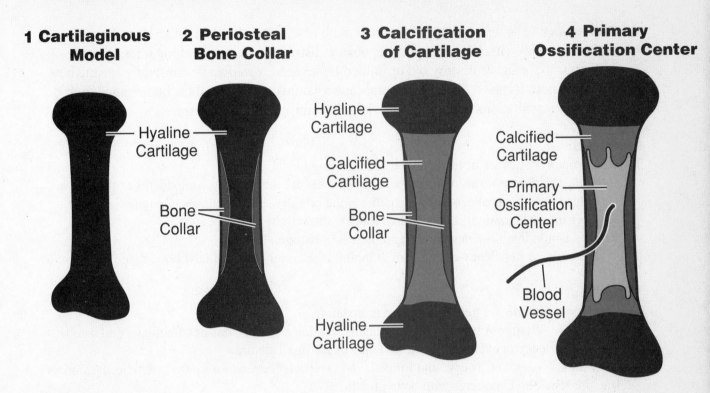

1 Cartilaginous Model

Hyaline Cartilage

2 Periosteal Bone Collar

Hyaline Cartilage

Bone Collar

3 Calcification of Cartilage

Hyaline Cartilage

Calcified Cartilage

Bone Collar

Hyaline Cartilage

4 Primary Ossification Center

Calcified Cartilage

Primary Ossification Center

Blood Vessel

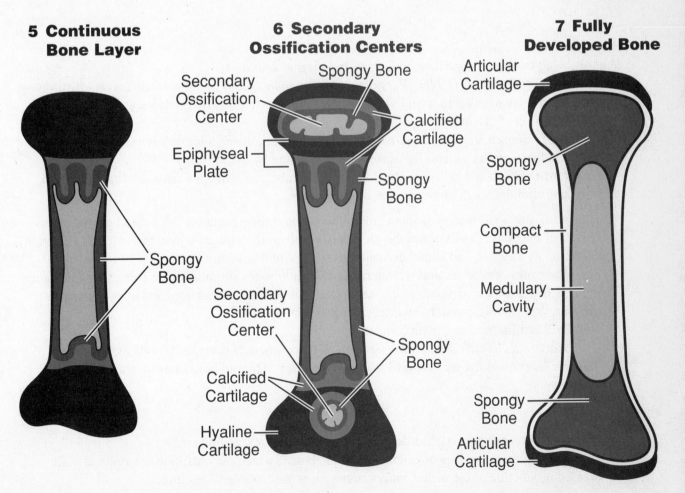

5 Continuous Bone Layer

Spongy Bone

6 Secondary Ossification Centers

Secondary Ossification Center

Spongy Bone

Epiphyseal Plate

Calcified Cartilage

Spongy Bone

Secondary Ossification Center

Calcified Cartilage

Spongy Bone

Hyaline Cartilage

7 Fully Developed Bone

Articular Cartilage

Spongy Bone

Compact Bone

Medullary Cavity

Spongy Bone

Articular Cartilage

SKELETAL SYSTEM / Bone Homeostasis

Remodeling   The replacement of old bone with new bone is called remodeling. Bone tissue constantly renews itself throughout life; bone redistributes its matrix along lines of mechanical stress. *Osteoclasts* destroy old or injured bone tissue; *osteoblasts* constantly secrete new bone matrix in its place. Certain minerals and vitamins are essential for bone remodeling. Mineral concentrations in the blood are largely controlled by hormones.

Minerals
The following minerals are needed for bone remodeling :
Calcium and Phosphorus   Calcium and phosphorus are the primary components of tricalcium phosphate (hydroxyapatite); hydroxyapatite is the primary salt that makes bone matrix hard.
Magnesium   Magnesium is needed to stimulate the activity of osteoclasts.
Boron   Boron inhibits calcium loss and increases estrogen levels.
Manganese   A deficiency of manganese inhibits the laying down of new bone tissue.

Vitamins
The following vitamins are needed for bone remodeling :
Vitamin A   Vitamin A helps to control the distribution and activities of osteoblasts and osteoclasts; deficiency or excess (toxic doses) may cause small stature.
Vitamin C   Vitamin C is essential for collagen synthesis (needed for matrix); deficiency hinders fracture repair and interferes with bone growth.

Hormones
The following hormones contribute to normal bone tissue activity :
Parathyroid Hormone (PTH)   Parathyroid hormone *increases* blood calcium levels. It stimulates the release of calcium from bone, inhibits calcium excretion by the kidneys, and stimulates the absorption of calcium by the intestine.
Calcitriol (vitamin D)   Calcitriol *increases* blood calcium levels. It inhibits calcium excretion by the kidneys and stimulates the absorption of calcium by the intestine.
Calcitonin   Calcitonin *decreases* blood calcium levels. It inhibits the release of calcium from bone and stimulates calcium excretion by the kidneys.

Human Growth Hormone   Human growth hormone is reponsible for the general growth of all tissues of the body. It stimulates the growth of epiphyseal cartilage, stimulates calcium excretion excretion by kidneys, and stimulates calcium absorption by the intestine.
Sex Hormones (estrogens and testosterone)   Sex hormones stimulate bone formation and cause the degeneration of cartilage in epiphyseal plates (stops longitudinal bone growth). Estrogens prevent osteoporosis, possibly by direct effect on osteoblasts.
Insulin   Insulin increases bone formation.
Glucocorticoids (cortisol and aldosterone)   Glucocorticoids decrease blood calcium levels.
Thyroid Hormones (thyroxine and triiodothyronine)   Thyroid hormones increase blood calcium levels.

Aging
Osteoporosis : age-related disorder characterized by decreased bone mass.
During aging there is a loss of calcium from the bones, which may result in osteoporosis. Decreased production of the matrix makes bones more susceptible to fracture.

CALCIUM BALANCE
Calcium homeostasis is regulated by Parathyroid Hormone (PTH), Calcitriol, and Calcitonin

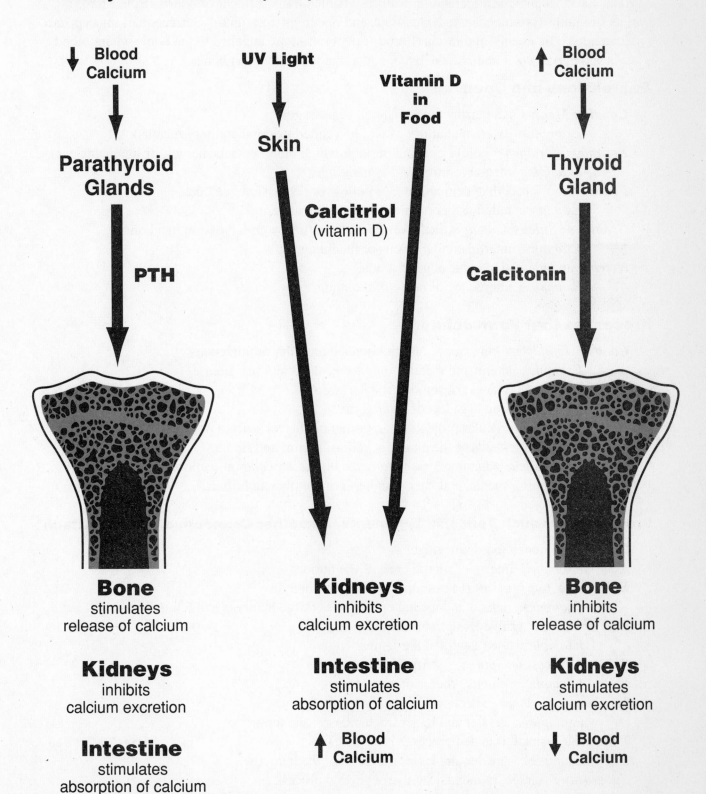

↓ Blood Calcium

UV Light

Vitamin D in Food

↑ Blood Calcium

Parathyroid Glands

Skin

Thyroid Gland

Calcitriol (vitamin D)

PTH

Calcitonin

Bone
stimulates release of calcium

Kidneys
inhibits calcium excretion

Bone
inhibits release of calcium

Kidneys
inhibits calcium excretion

Intestine
stimulates absorption of calcium

Kidneys
stimulates calcium excretion

Intestine
stimulates absorption of calcium

↑ Blood Calcium

↓ Blood Calcium

↑ Blood Calcium

SKELETAL SYSTEM / Bone Markings

Surface Markings

Surface markings include all structural features visible on the surfaces of bones. Bones have a variety of bumps (prominences), depressions, and openings (foramina). Surface markings appear where tendons, ligaments, and fascia (fibrous membranes) are attached to the bone; where blood vessels and nerves enter and exit the bone; and at the joints between bones.

Depressions and Openings

Canal or Meatus : a narrow tube, channel, or passageway.
  Example : external auditory canal (also called external auditory meatus).
Foramen (*foramen* = hole) *:* opening through which blood vessels, nerves, or ligaments pass.
  Example : mental foramen of the mandible.
Fossa (*fossa* = basinlike depression) *:* a hollow or depression in a bone.
  Example : radial and coronoid fossae of the humerus.
Groove or Sulcus (*sulcus* = ditchlike groove) *:* a furrow or depression in a bone.
  Example : intertubercular sulcus of the humerus.
Notch : an indentation at the edge of a bone.
  Example : intercondylar notch of the femur.

Processes that Form Joints

Condyle (*condyle* = knuckle) *:* a large, rounded articular prominence.
  Example : lateral and medial condyles of the femur and tibia.
Facet (*facet* = little face) *:* a smooth, flat surface.
  Example : articular facet of a vertebra.
Head : a rounded articular projection supported on the constricted portion of a bone.
  Examples : heads of the humerus, radius, femur, and fibula.
Malleolus (*malleus* = hammer) *:* a hammerhead-like articular projection.
  Examples : medial and lateral malleoli of the tibia and fibula.

Processes to which Tendons, Ligaments, and other Connective Tissues Attach

Crest : a prominent border or ridge.
  Example : intertrochanteric crest of the femur.
Epicondyle (*epi* = upon) *:* a prominence above a condyle.
  Examples : lateral and medial epicondyles of the humerus and femur.
Line or Linea : a ridge less prominent than a crest.
  Example : linea aspera of the femur.
Spinous Process or Spine : a sharp, slender process.
  Example : spinous process of a vertebra.
Trochanter : a large projection found only on the femur.
  Examples : greater and lesser trochanters of the femur.
Tubercle : a small, rounded process.
  Examples : greater and lesser tubercles of the humerus.
Tuberosity : a large, rounded, usually roughened process.
  Examples : deltoid, radial, and gluteal tuberosities of the humerus, radius, and femur.

BONE MARKINGS

Upper Extremity
(anterior view)

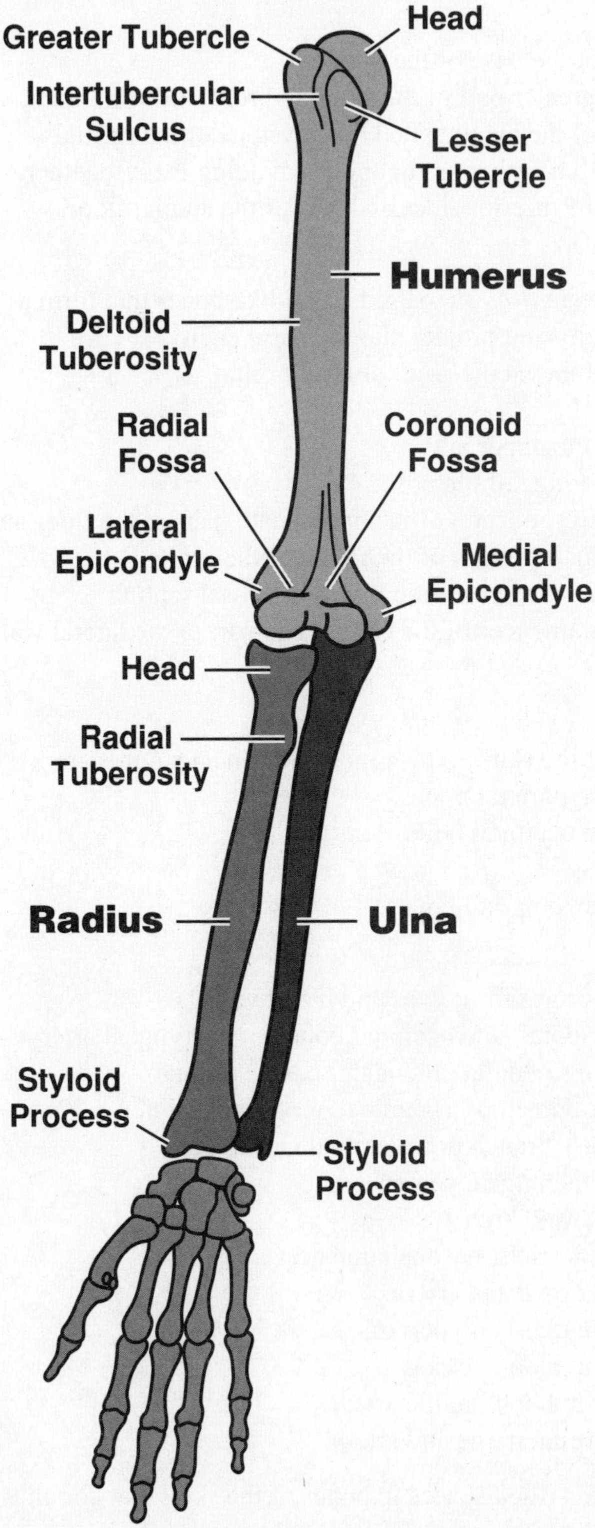

Greater Tubercle

Head

Intertubercular Sulcus

Lesser Tubercle

Humerus

Deltoid Tuberosity

Radial Fossa

Coronoid Fossa

Lateral Epicondyle

Medial Epicondyle

Head

Radial Tuberosity

Radius

Ulna

Styloid Process

Styloid Process

Lower Extremity
(posterior view)

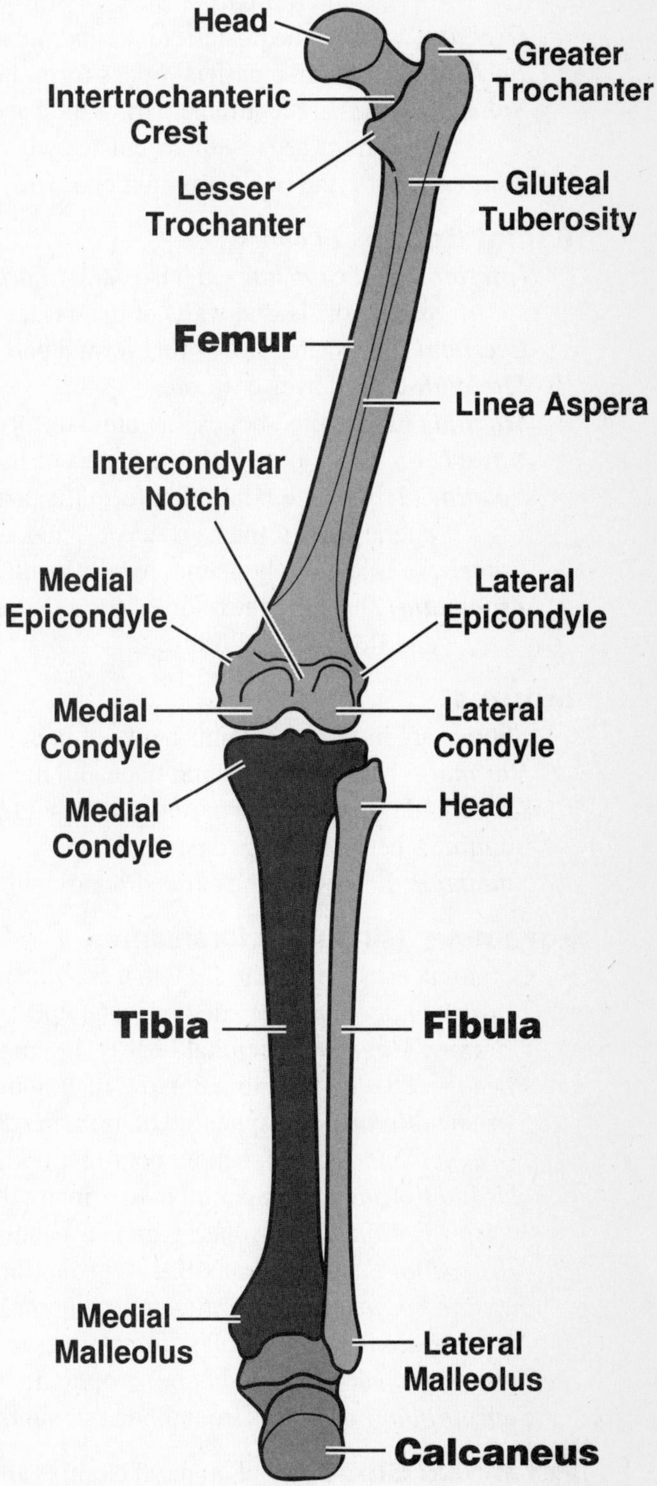

Head

Greater Trochanter

Intertrochanteric Crest

Lesser Trochanter

Gluteal Tuberosity

Femur

Linea Aspera

Intercondylar Notch

Medial Epicondyle

Lateral Epicondyle

Medial Condyle

Lateral Condyle

Medial Condyle

Head

Tibia

Fibula

Medial Malleolus

Lateral Malleolus

Calcaneus

SKELETAL SYSTEM / Skull

Cranial Bones   *8 bones*
Ethmoid : located in the anterior part of the cranial floor between the orbits (eye sockets); anterior to the sphenoid and posterior to the nasal bones.

Frontal : forms the anterior part of the cranium (forehead), the roofs of the orbits, and most of the anterior part of the cranial floor.

Occipital : forms the posterior part and most of the base of the cranium.

Parietal (2) : the two parietal bones form the greater part of the sides and roof of the cranium.

Sphenoid : lies at the middle part of the base of the skull; called the keystone of the cranial floor because it articulates with all the other cranial bones, holding them together.

Temporal (2) : the two temporal bones form the inferior sides and part of the cranial floor.

Facial Bones   *14 bones*
Inferior Nasal Conchae (2) also called *Turbinate Bones* : paired scroll-like bones that form a part of the lateral walls of the nasal cavity and project into the nasal cavity.

Lacrimal (2) : paired bones that form a part of the medial wall of each orbit.

Mandible : the lower jaw bone.

Maxillae (2) : paired bones that unite to form the upper jaw.

Nasal (2) : paired bones that form part of the bridge of the nose.

Palatine (2) : paired bones that form the posterior portion of the hard palate, part of the floor and lateral wall of the nasal cavity, and a small portion of the floors of the orbits.

Vomer : a triangular bone that forms the inferior and posterior part of the nasal septum.

Zygomatic (2) : the cheek bones; form the prominences of the cheeks and part of the lateral wall and floor of each orbit.

Sutures
Sutures are immovable joints between bones of the skull. The 4 most prominent sutures are :

Coronal : between the frontal bone and the two parietal bones.

Lambdoid : between the parietal bones and the occipital bone.

Sagittal : between the two parietal bones.

Squamous : between the parietal bones and the temporal bones.

Foramina   (singular : foramen)
Foramina are holes in the skull that provide passages for nerves and blood vessels.

Foramen Lacerum (bounded by sphenoid, temporal, and occipital bones) : pharyngeal artery.

Foramen Magnum (occipital bone) : brainstem (medulla oblongata) passes through.

Foramen Ovale (sphenoid bone) : mandibular branch of trigeminal nerve (cranial nerve V).

Foramen Rotundum (sphenoid bone) : maxillary branch of trigeminal nerve.

Foramen Spinosum (sphenoid bone) : middle meningeal vessels.

Carotid Foramen (temporal bone) : internal carotid artery.

Incisive Foramen (maxillae) : greater palatine vessels; nasopalatine nerve.

Infraorbital Foramen (maxillae) : infraorbital nerve and artery.

Jugular Foramen (temporal bone) : internal jugular vein; nerves.

Mental Foramen (mandible) : mental nerve and blood vessels.

Optic Foramen (sphenoid bone) : optic nerve and ophthalmic artery.

Supraorbital Foramen (frontal bone) : supraorbital nerve and artery.

Paranasal Sinuses   Paranasal cavities are air-filled spaces in bones of the skull that communicate with the nasal cavity and are lined by mucous membranes. The cranial bones containing paranasal sinuses are the frontal, sphenoid, ethmoid, and maxillae.

SKULL : lateral view

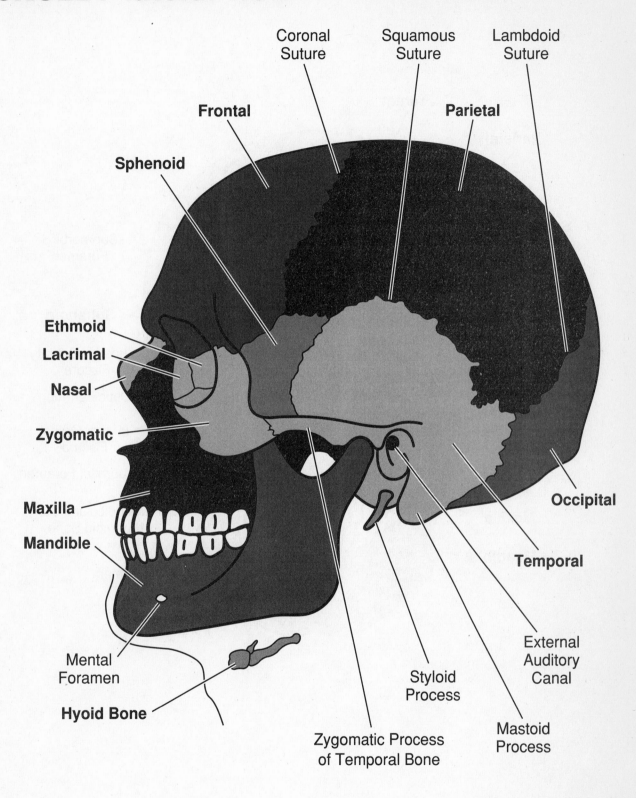

Coronal Suture

Squamous Suture

Lambdoid Suture

Frontal

Parietal

Sphenoid

Ethmoid

Lacrimal

Nasal

Zygomatic

Occipital

Maxilla

Mandible

Temporal

Mental Foramen

External Auditory Canal

Hyoid Bone

Styloid Process

Mastoid Process

Zygomatic Process of Temporal Bone

SKULL : anterior view

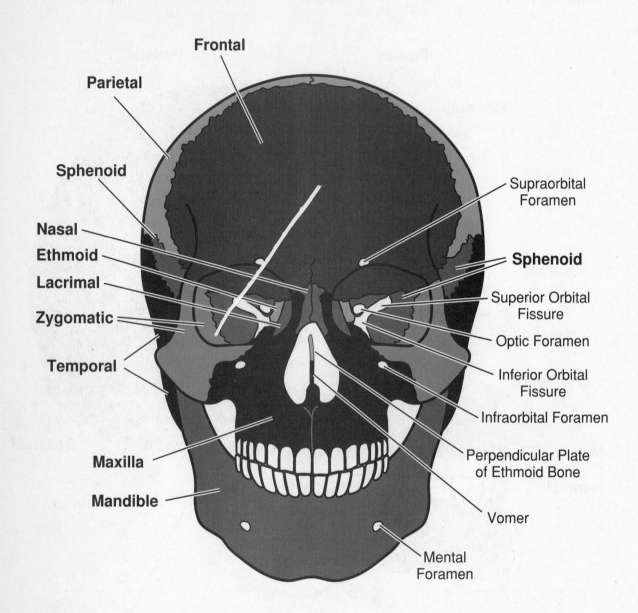

Frontal

Parietal

Sphenoid

Nasal

Ethmoid

Lacrimal

Zygomatic

Temporal

Maxilla

Mandible

Supraorbital Foramen

Sphenoid

Superior Orbital Fissure

Optic Foramen

Inferior Orbital Fissure

Infraorbital Foramen

Perpendicular Plate of Ethmoid Bone

Vomer

Mental Foramen

SKULL : posterior view

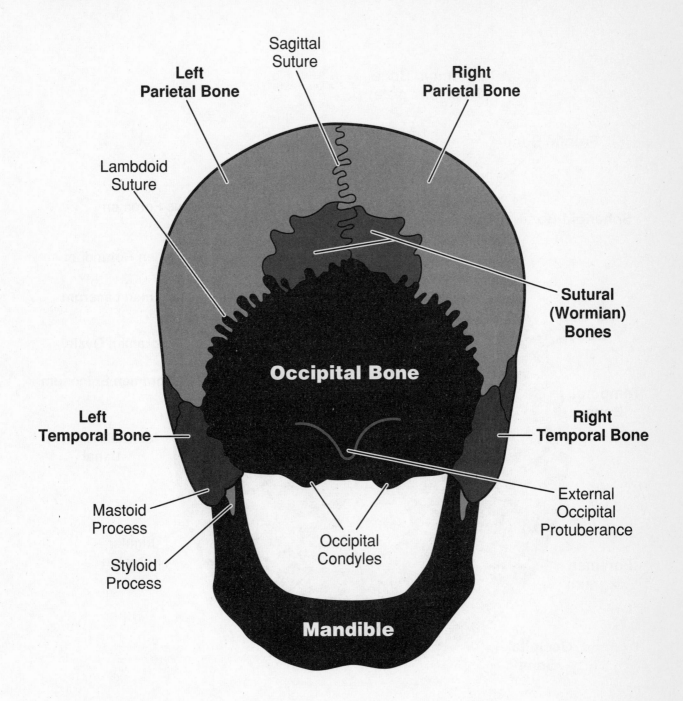

Sagittal
Suture

**Left
Parietal Bone**

**Right
Parietal Bone**

Lambdoid
Suture

Sutural
(Wormian)
Bones

Occipital Bone

**Left
Temporal Bone**

**Right
Temporal Bone**

Mastoid
Process

External
Occipital
Protuberance

Styloid
Process

Occipital
Condyles

Mandible

SKULL : floor of the cranium (from above)

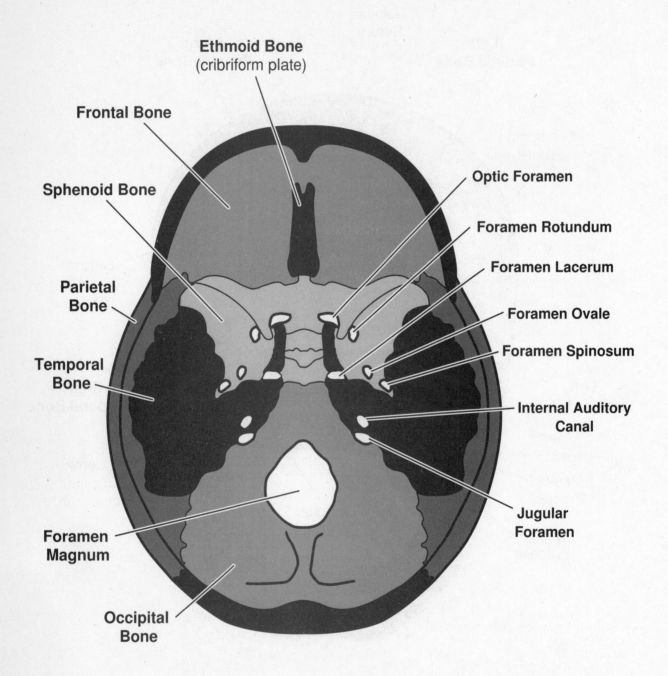

Ethmoid Bone
(cribriform plate)

Frontal Bone

Sphenoid Bone

Optic Foramen

Foramen Rotundum

Foramen Lacerum

Parietal
Bone

Foramen Ovale

Foramen Spinosum

Temporal
Bone

Internal Auditory
Canal

Foramen
Magnum

Jugular
Foramen

Occipital
Bone

SKULL : inferior view

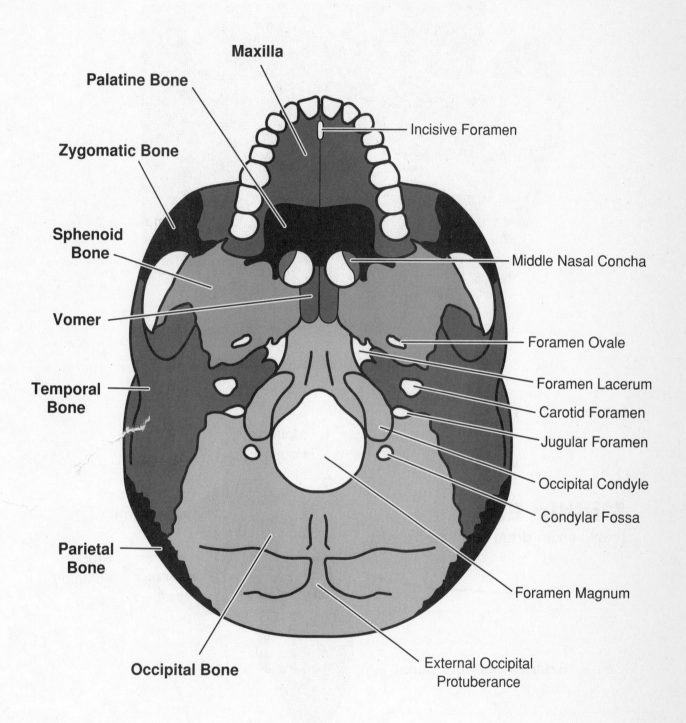

Maxilla

Palatine Bone

Zygomatic Bone

Incisive Foramen

Sphenoid
Bone

Middle Nasal Concha

Vomer

Foramen Ovale

Temporal
Bone

Foramen Lacerum

Carotid Foramen

Jugular Foramen

Occipital Condyle

Condylar Fossa

Parietal
Bone

Foramen Magnum

Occipital Bone

External Occipital
Protuberance

AUDITORY OSSICLES

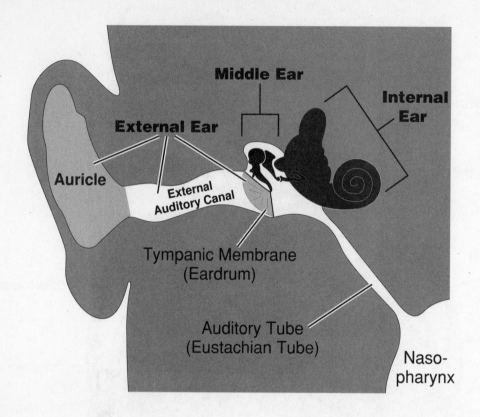

Middle Ear

External Ear

Internal Ear

Auricle

External Auditory Canal

Tympanic Membrane (Eardrum)

Auditory Tube (Eustachian Tube)

Naso-pharynx

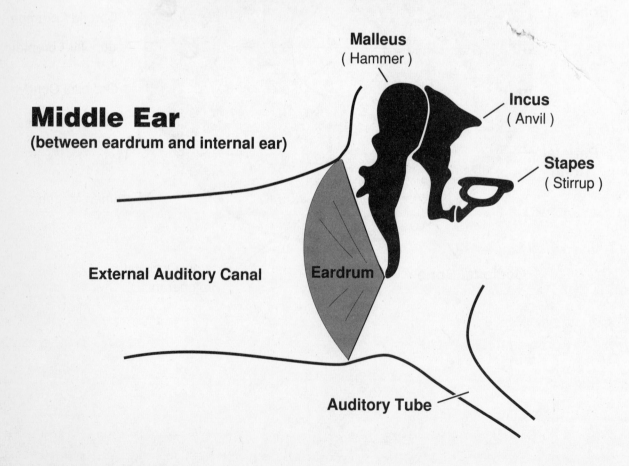

Middle Ear
(between eardrum and internal ear)

Malleus (Hammer)

Incus (Anvil)

Stapes (Stirrup)

External Auditory Canal

Eardrum

Auditory Tube

PARANASAL SINUSES

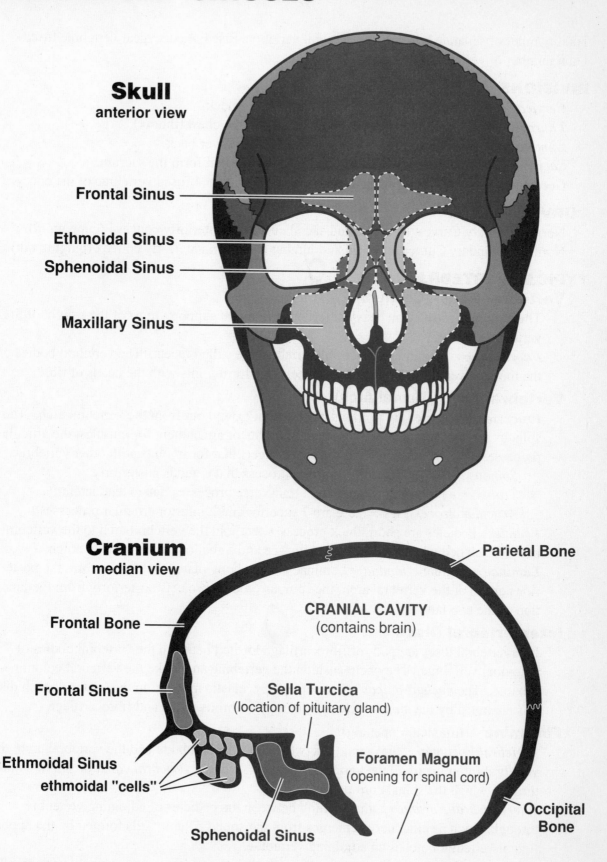

Skull
anterior view

Frontal Sinus

Ethmoidal Sinus

Sphenoidal Sinus

Maxillary Sinus

Cranium
median view

Parietal Bone

CRANIAL CAVITY
(contains brain)

Frontal Bone

Frontal Sinus

Sella Turcica
(location of pituitary gland)

Ethmoidal Sinus

ethmoidal "cells"

Foramen Magnum
(opening for spinal cord)

Occipital
Bone

Sphenoidal Sinus

note : maxillary sinus is not visible in this illustration

23

SKELETAL SYSTEM / Vertebral Column (Spine)

Total number of separate bones = 26 (5 sacral vertebrae fused; 4 coccygeal vertebrae fused).
Total number of vertebrae = 33.

DIVISIONS

Cervical (cervix = neck) : 7 vertebrae located in the neck.

Thoracic (thorax = chest) : 12 vertebrae located in the chest (thorax).

Lumbar (lumbus = loin) : 5 vertebrae located in the lower back.

Sacral (os sacrum = sacred bone) : 5 fused vertebrae that form the sacrum.

Coccygeal (coccyx = bone shaped like a cuckoo's beak) : 4 fused vertebrae of the coccyx.

CURVES

Normal Primary Curves (thoracic and sacral curves) : anteriorly concave (cupping in).

Normal Secondary Curves (cervical and lumbar curves) : anteriorly convex (bulging out).

TYPICAL VERTEBRA
Vertebral Body (Centrum)

The vertebral body is the heavy anterior portion that supports most of the weight of the vertebral column.

Facets   The articulating surfaces of vertebrae are called facets. The vertebral bodies of the thoracic vertebrae have articular facets that form joints with the heads of ribs.

Vertebral Arch (Neural Arch)

Processes   Processes are portions of bone that extend out from the vertebral arch. The spinous and transverse processes serve as points of attachment for muscles; the articular processes have articulating surfaces called facets that form joints with other vertebrae.

> *Spinous Process :* there is 1 spinous process that extends posteriorly.

> *Transverse Processes :* there are 2 transverse processes that extend laterally.

> *Articular Processeses :* there are 2 superior and 2 inferior articular processes.

Pedicles   Pedicles are short, thick processes that join the vertebral arch to the vertebral body. Each pedicle has a deep lower surface and a shallow upper surface notch.

Laminae (singular : *lamina*) Laminae are flat bony plates that join to form the posterior portion of the vertebral arch; the spinous process projects posteriorly from the junction of the two laminae.

Intervertebral Discs

Intervertebral discs are pads of fibrocartilage located between the vertebral bodies of adjoining vertebrae. They help cushion the vertebrae and make the vertebral column flexible. The *nucleus pulposus* is a soft, pulpy, elastic material in the center of each disc; it is encircled by the *annulus fibrosus*, a ring of fibrous tissue and fibrocartilage.

Foramina   (foramen = opening)

Vertebral Foramen : the opening between the vertebral body and the vertebral arch; the vertebral foramina of all the vertebrae together form the *vertebral canal* or *spinal canal* through which the spinal cord passes.

Intervertebral Foramen : the opening between the pedicles of adjoining vertebrae through which a spinal nerve emerges from the spinal column; it is formed by the apposition of vertebral notches on adjoining vertebrae.

Transverse Foramen : the opening in each transverse process of a *cervical* vertebra; the vertebral artery, vertebral vein, and nerve fibers pass through the transverse foramen.

VERTEBRAL COLUMN : 3 Views

Anterior **Lateral** **Posterior**

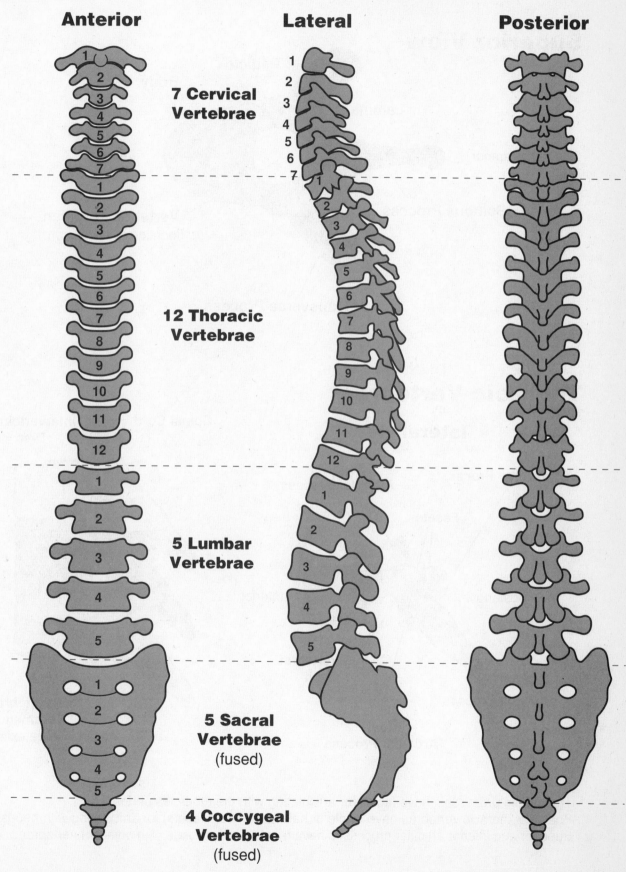

7 Cervical
Vertebrae

12 Thoracic
Vertebrae

5 Lumbar
Vertebrae

5 Sacral
Vertebrae
(fused)

4 Coccygeal
Vertebrae
(fused)

TYPICAL VERTEBRA

Superior View

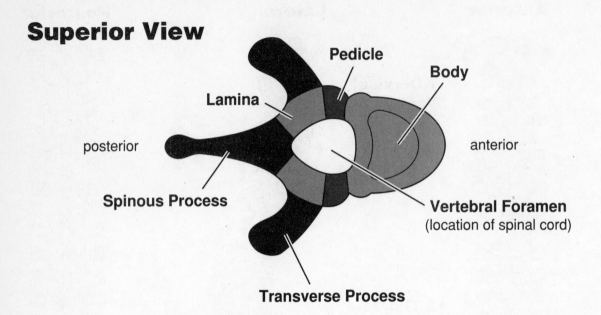

Pedicle

Body

Lamina

posterior

anterior

Spinous Process

Vertebral Foramen
(location of spinal cord)

Transverse Process

Thoracic Vertebra

lateral view

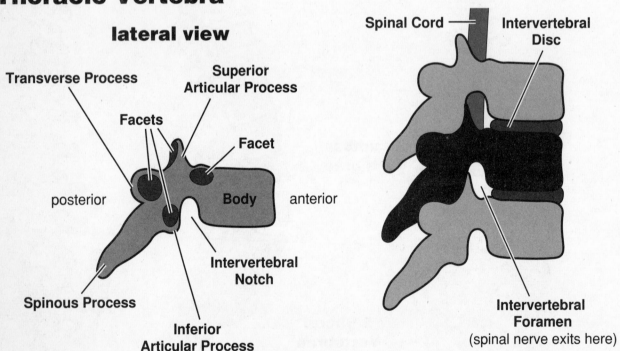

Transverse Process

Superior
Articular Process

Facets

Facet

posterior

Body

anterior

Intervertebral
Notch

Spinous Process

Inferior
Articular Process

Spinal Cord

Intervertebral
Disc

Intervertebral
Foramen
(spinal nerve exits here)

Facets :

Transverse processes have facets for articulating with the tubercles of ribs.
Bodies of thoracic vertebrae have whole or half-facets (demifacets) for articulating with heads of ribs.
Superior and inferior articular processes have facets that articulate with adjacent vertebrae.

VERTEBRAL SHAPES
Superior View

Cervical Vertebra
(Atlas / 1st Cervical Vertebra)

posterior

Posterior Arch

Vertebral Foramen

Transverse Process

Superior Articular Facet

anterior

Anterior Arch

Transverse Foramen
(location of vertebral artery)

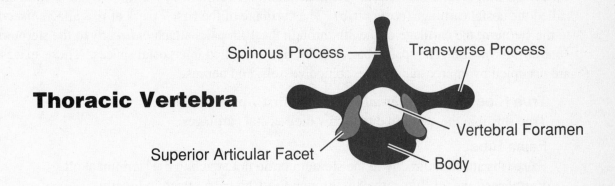

Thoracic Vertebra

Spinous Process

Transverse Process

Superior Articular Facet

Vertebral Foramen

Body

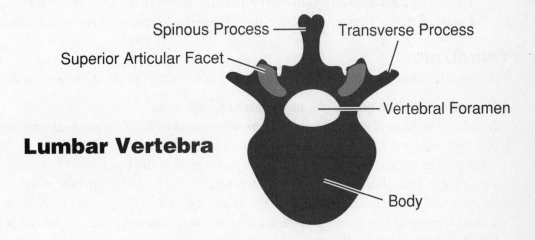

Lumbar Vertebra

Spinous Process

Transverse Process

Superior Articular Facet

Vertebral Foramen

Body

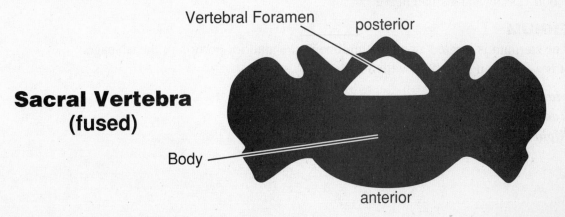

Sacral Vertebra
(fused)

Vertebral Foramen

posterior

Body

anterior

SKELETAL SYSTEM / Thoracic Cage (Rib Cage)

The term *thorax* refers to the entire chest. The skeletal portion of the thorax is called the *thoracic cage* or *rib cage*; it is formed by the ribs, thoracic vertebrae, sternum, and costal cartilages.

Structures :   12 pairs of ribs, 12 thoracic vertebrae, sternum (breastbone), and costal cartilages.

Functions
   Support : support of the arms and pectoral girdles (shoulder girdles).
   Protection : protection of the visceral organs in the thoracic and upper abodominal cavities.
   Ventilation : ventilation of the lungs (inspiration and expiration of air).

RIBS
   There are 12 pairs of ribs; one attached to each of the 12 thoracic vertebrae.
   Costal Cartilage   The first ten ribs are attached to the sternum by a strip of hyaline cartilage called the costal cartilage (*costa* = rib). The cartilage of the first 7 pairs of ribs attach *directly* to the sternum; the cartilage of the 8th through the 10th pairs attach *indirectly* to the sternum.
   Intercostal Spaces   The spaces between the ribs are called intercostal spaces. These spaces are occupied by intercostal muscles, blood vessels, and nerves.

True Ribs *(Vertebrosternal Ribs) :* the first 7 pairs of ribs.
   True ribs join the sternum directly by their costal cartilages.

False Ribs
   False ribs attach indirectly to the sternum or do not attach to the sternum at all.
   Vertebrochondral Ribs : the 8th through the 10th pairs attach indirectly.
   The costal cartilages attach to each other and then to the cartilages of the 7th pair of ribs.
   Floating Ribs (Vertebral Ribs) : the 11th and 12th pairs do not attach to the sternum.

TYPICAL RIB
   A typical rib has a long, slender shaft that curves around the chest and slopes downward.

   Body (Shaft) : The body is the main part of the rib.
   Head :   The head is a projection at the posterior end of a rib. It consists of one or two facets that articulate with facets on the bodies of adjacent thoracic vertebrae.
   Neck :   The neck is a constricted portion just lateral to the head.
   Tubercle : The tubercle is a knoblike structure on the posterior surface where the neck joins the body. A tubercle articulates with the facet of a transverse process on a thoracic vertebra.
   Costal Groove : The costal groove is a groove on the inner surface of the rib; it protects blood vessels and a small nerve.

STERNUM
   The sternum is located along the midline in the anterior portion of the rib cage.
   It is a flat, elongated bone with 3 parts.

   Manubrium : the upper portion.
   Body : the middle, largest portion of the sternum.
   Xiphoid Process : the lowest portion of the sternum.

THORACIC CAGE

True Ribs (1 — 7) : attach directly to the sternum by costal cartilages.
False Ribs (8 — 12) : attach to the sternum indirectly or not at all.
Floating Ribs (11 — 12) : do not attach to the sternum.

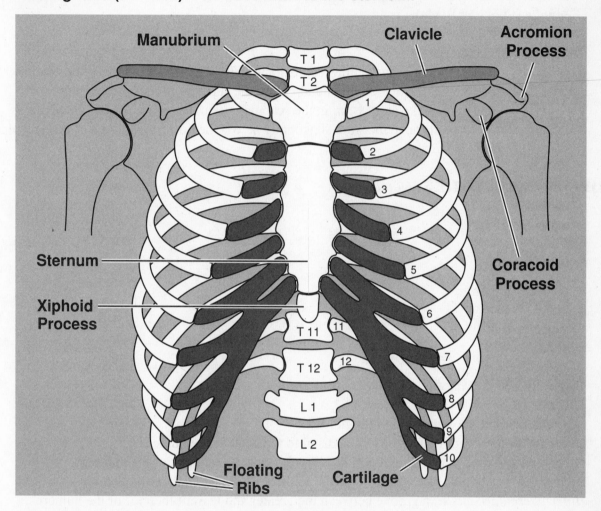

Thoracic Vertebra (viewed from above)

All ribs articulate with their respective thoracic vertebrae.

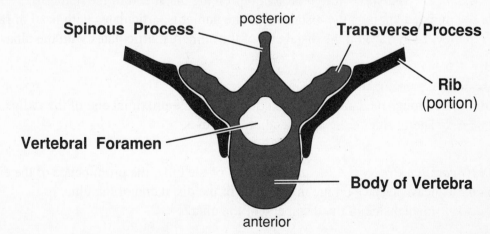

PECTORAL GIRDLES (Shoulder Girdles) *4 bones*
Clavicle (Collarbone) (2)
Scapula (Shoulder Blade) (2)

Special Structures
Acromion Process : superior end of the scapular spine; most prominent point of the shoulder.
Coracoid Process : a projection at the lateral end of the superior border of the scapula.
Glenoid Cavity : socket-like depression in the scapula; articulates with head of the humerus.
Scapular Spine : a sharp ridge that runs diagonally across the posterior surface of the scapula.

UPPER EXTREMITIES *60 bones*
Humerus (2)
Radius (2)
Ulna (2)
Carpals (16)

Capitate (2)	Lunate (2)	Scaphoid (2)	Trapezoid (2)
Hamate (2)	Pisiform (2)	Trapezium (2)	Triquetrum (2)

Metacarpals (10)
Phalanges (28)

Special Structures (from proximal to distal locations)
Humerus
Head : enlarged, proximal end of the humerus.
Anatomical Neck : oblique groove just distal to the head of the humerus.
Greater Tubercle : lateral projection distal to the anatomical neck of the humerus.
Lesser Tubercle : anterior projection distal to the anatomical neck of the humerus.
Intertubercular Sulcus : a groove between the greater and lesser tubercles of the humerus.
Surgical Neck : constricted portion of the humerus just distal to the tubercles.
Deltoid Tuberosity : roughened, V-shaped area; the lateral, middle portion of the humerus.
Radial and *Coronoid Fossae :* depressions on the anterior side of the distal end of the humerus.
Lateral & Medial Epicondyles : rough projections on the distal end of the humerus.
Capitulum : a rounded knob on the distal end of the humerus; articulates with head of radius.
Trochlea : pulleylike surface on the distal end of the humerus; articulates with the ulna.

Radius
Head : disc-shaped proximal end of the radius.
Radial Tuberosity : roughened area on the medial side of the proximal end of the radius.
Styloid Process : lateral side of the distal end of the radius.

Ulna
Olecranon (Olecranon Process) : the proximal end of the ulna; the prominence of the elbow.
Ulnar Notch : concave surface on the medial side of the distal end of the ulna.
Styloid Process : medial side of the distal end of the ulna.

PECTORAL GIRDLE and UPPER EXTREMITY

Anterior View

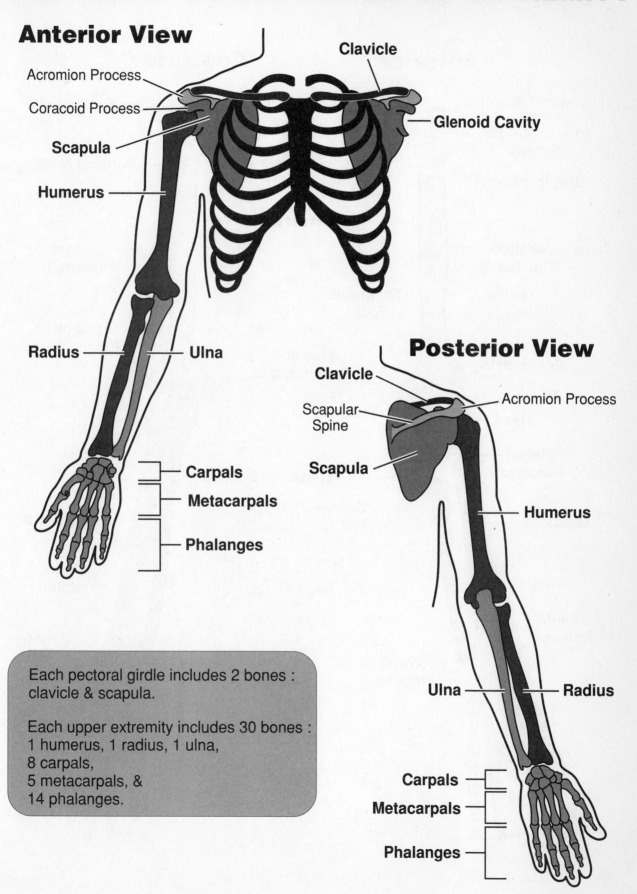

Clavicle

Acromion Process

Coracoid Process

Glenoid Cavity

Scapula

Humerus

Radius

Ulna

Posterior View

Clavicle

Scapular Spine

Acromion Process

Scapula

Humerus

Ulna

Radius

Carpals

Metacarpals

Phalanges

Carpals

Metacarpals

Phalanges

Each pectoral girdle includes 2 bones : clavicle & scapula.

Each upper extremity includes 30 bones :
1 humerus, 1 radius, 1 ulna,
8 carpals,
5 metacarpals, &
14 phalanges.

UPPER EXTREMITY

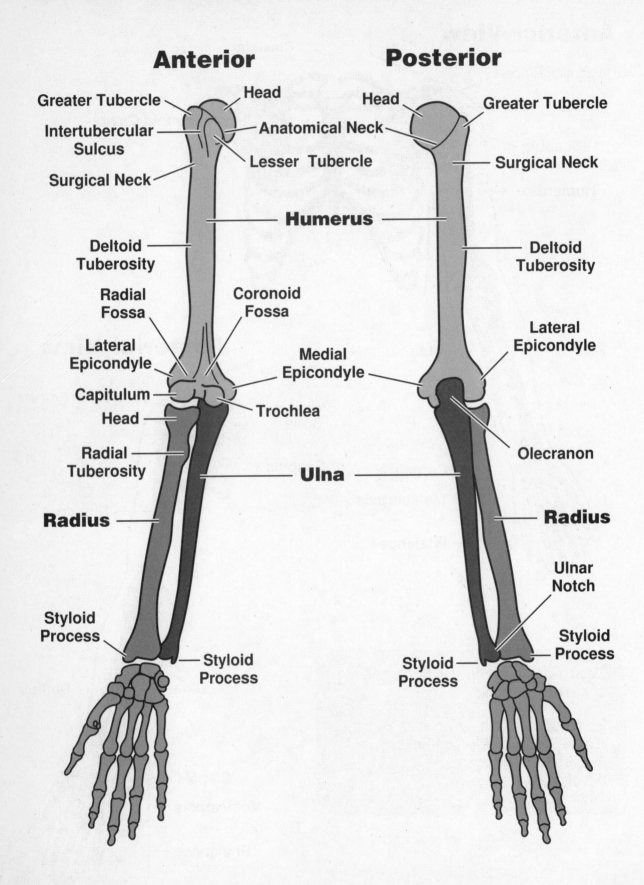

Anterior

Posterior

Greater Tubercle

Head

Head

Greater Tubercle

Intertubercular Sulcus

Anatomical Neck

Surgical Neck

Lesser Tubercle

Surgical Neck

Humerus

Deltoid Tuberosity

Deltoid Tuberosity

Radial Fossa

Coronoid Fossa

Lateral Epicondyle

Lateral Epicondyle

Medial Epicondyle

Capitulum

Trochlea

Head

Olecranon

Radial Tuberosity

Ulna

Radius

Radius

Ulnar Notch

Styloid Process

Styloid Process

Styloid Process

Styloid Process

32

RIGHT HAND

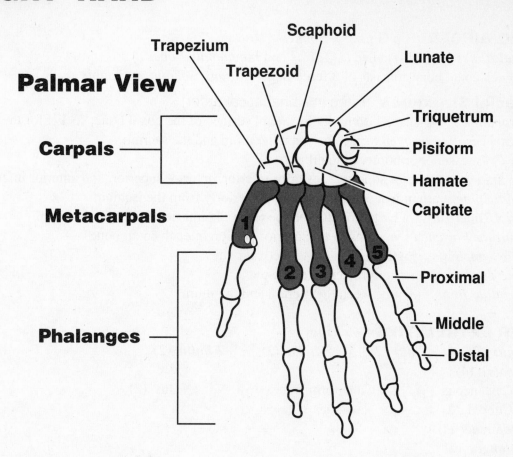

Palmar View

Trapezium
Trapezoid
Scaphoid
Lunate
Triquetrum
Pisiform
Hamate
Capitate

Carpals

Metacarpals

1 2 3 4 5

Phalanges

Proximal
Middle
Distal

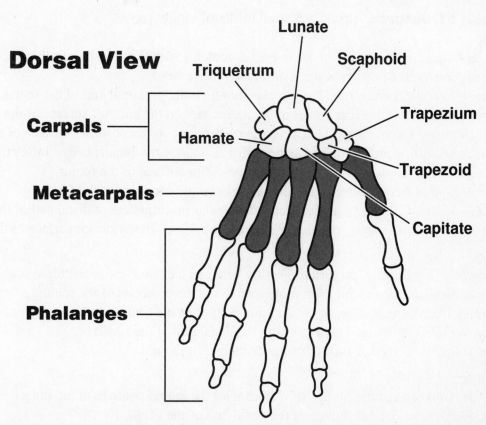

Dorsal View

Lunate
Triquetrum
Scaphoid
Hamate
Trapezium
Trapezoid
Capitate

Carpals

Metacarpals

Phalanges

PELVIC GIRDLE (Hip Girdle)   *2 bones*

Coxal Bones (2)   also called *Os Coxae* and *Innominate Bones*
Each coxal bone consists of 3 fused bones : *ilium, ischium, and pubis.*

Special Structures (listed in alphabetical order)

Acetabulum : cup-shaped cavity in the lateral surface of the coxal bone; socket for the femur.
Greater Sciatic Foramen : space between the ilium and the sacrum.
Iliac Crest : superior border of the ilium.
Iliac Spines : posterior superior, posterior inferior, anterior superior, and anterior inferior.
Ischial Spine : prominent spine projecting posteriorly from the ischium.
Ischial Tuberosity : a large, rough prominence at the inferior end of the ischium.
Obturator Foramen : opening in the inferior portion of each coxal bone.
Pubic Symphysis : joint between the two coxal bones.
Pubic Tubercle : anterior projection of the pubis.
Sacroiliac Joint : joint between the sacrum and the ilium.

LOWER EXTREMITIES *60 bones*

Femur (2)   **Tibia** (2)   **Patella** (2)   **Fibula** (2)
Tarsals (14)
  Calcaneous (2)   Cuneiform Bones (6)   Talus (2)
  Cuboid (2)   Navicular (2)
Metatarsals (10)
Phalanges (28)

Special Structures (from proximal to distal locations)
Femur

Head : enlarged, proximal end of the femur; articulates with the acetabulum (hip socket).
Neck : the constricted portion of the femur distal to the head.
Greater & Lesser Trochanters : projections found on the proximal end of the femur.
Intertrochanteric Line : a line between the trochanters on the anterior surface of the femur.
Intertrochanteric Crest : a crest between the trochanters on the posterior surface of the femur.
Gluteal Tuberosity : projection on the posterior surface of the femur; below lesser trochanter.
Linea Aspera : a rough, vertical ridge on the posterior surface of the femur.
Lateral & Medial Epicondyles : prominences above the condyles of the femur.
Lateral & Medial Condyles : large, rounded articular prominences at distal end of the femur.
Intercondylar Fossa : depressed area between condyles on the posterior surface of the femur.

Tibia

Intercondylar Eminence : upward projection between the condyles of the tibia.
Lateral & Medial Epicondyles : prominences above the condyles of the femur.
Tibial Tuberosity : projection on the anterior surface of the tibia; patellar ligament attaches.
Fibular Notch : notch that articulates with the distal end of the fibula.
Medial Malleolus : medial surface of the distal end of the tibia.

Fibula

Head : proximal end of the fibula; articulates with the lateral condyle of the tibia.
Lateral Malleolus : lateral surface of the distal end of the fibula.

PELVIC GIRDLE and LOWER EXTREMITY

The pelvic girdle consists of 2 coxal bones.
Each coxal bone has 3 components : ilium, ischium, and pubis.

Each lower extremity includes 30 bones : 1 femur, 1 patella, 1 tibia,
1 fibula, 7 tarsals, 5 metatarsals, and 14 phalanges.

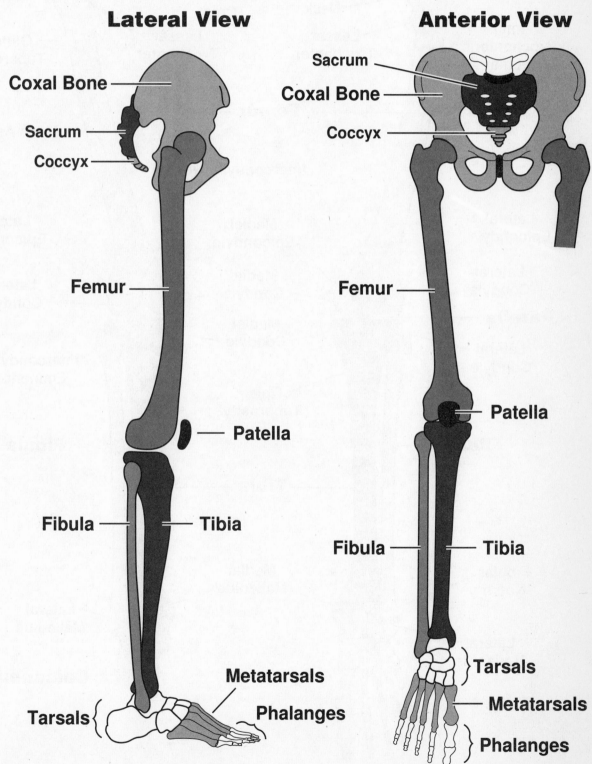

Lateral View

Coxal Bone

Sacrum

Coccyx

Femur

Patella

Fibula — Tibia

Tarsals

Metatarsals

Phalanges

Anterior View

Sacrum

Coxal Bone

Coccyx

Femur

Patella

Fibula — Tibia

Tarsals

Metatarsals

Phalanges

LOWER EXTREMITY

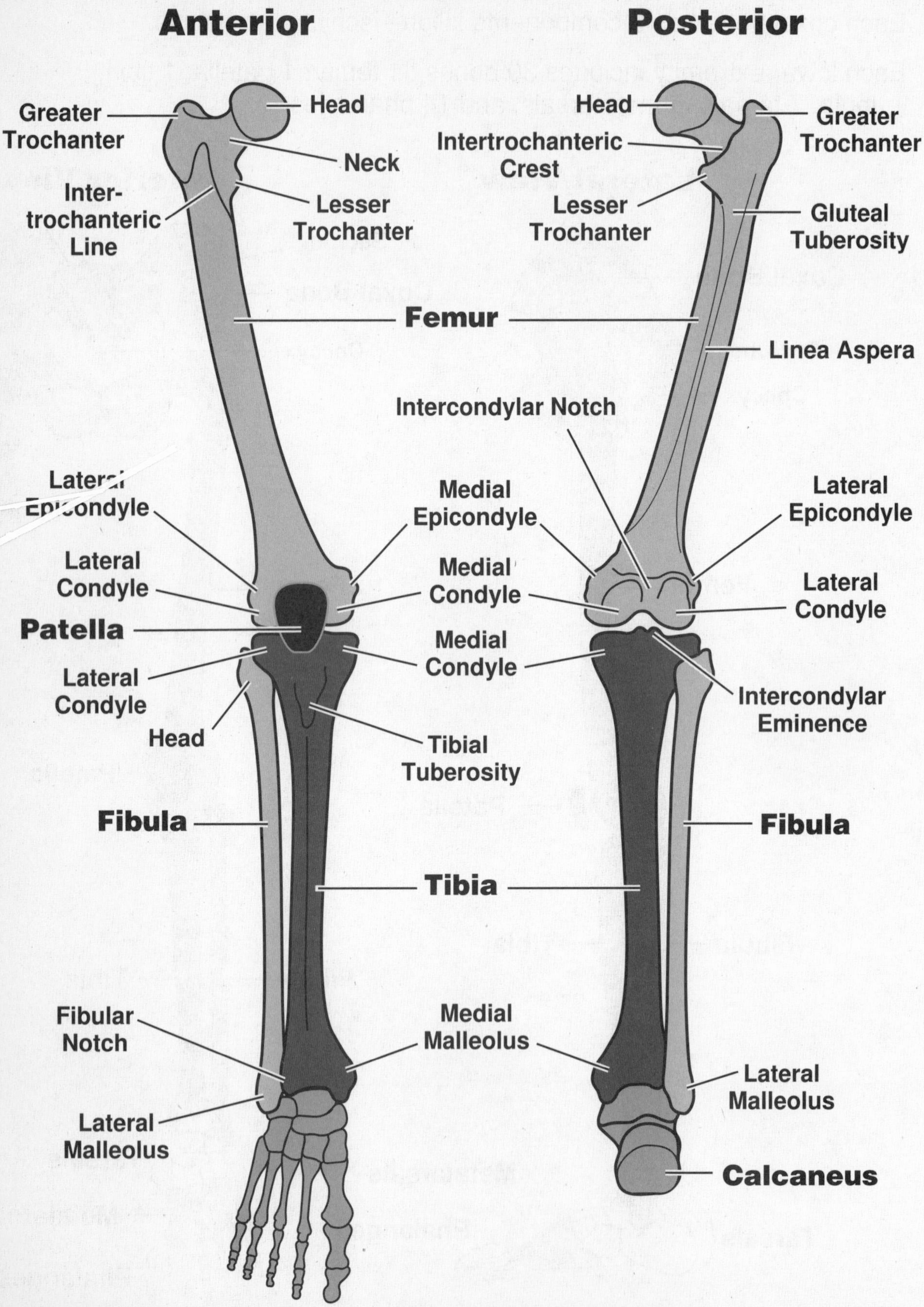

Anterior

Posterior

Greater Trochanter

Head

Inter-trochanteric Line

Neck

Lesser Trochanter

Femur

Lateral Epicondyle

Lateral Condyle

Patella

Lateral Condyle

Head

Fibula

Tibia

Fibular Notch

Lateral Malleolus

Medial Epicondyle

Medial Condyle

Medial Condyle

Tibial Tuberosity

Intercondylar Notch

Medial Malleolus

Head

Intertrochanteric Crest

Lesser Trochanter

Greater Trochanter

Gluteal Tuberosity

Linea Aspera

Lateral Epicondyle

Lateral Condyle

Intercondylar Eminence

Fibula

Lateral Malleolus

Calcaneus

RIGHT FOOT

Lateral View

Navicular

Talus

Intermediate Cuneiform

Metatarsals

Phalanges

Calcaneus

Cuboid

Lateral Cuneiform

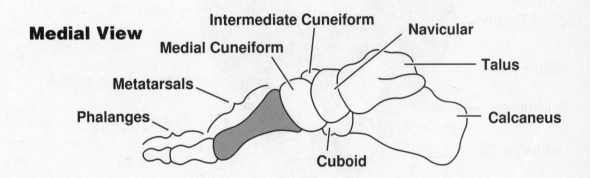

Medial View

Intermediate Cuneiform

Medial Cuneiform

Navicular

Talus

Metatarsals

Phalanges

Calcaneus

Cuboid

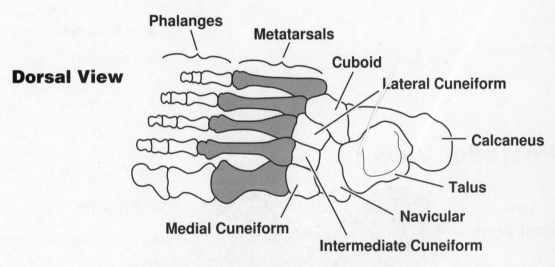

Dorsal View

Phalanges

Metatarsals

Cuboid

Lateral Cuneiform

Calcaneus

Talus

Navicular

Medial Cuneiform

Intermediate Cuneiform

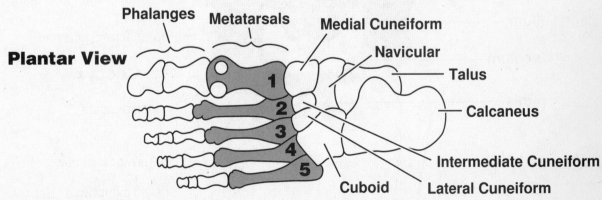

Plantar View

Phalanges

Metatarsals

Medial Cuneiform

Navicular

Talus

Calcaneus

Intermediate Cuneiform

Cuboid

Lateral Cuneiform

1

2

3

4

5

PELVIS

Pelvis : includes the pelvic girdle, sacrum, and coccyx.
Pelvic Girdle : consists of the 2 coxal bones.
Coxal Bone : has 3 fused components (ilium, ischium, and pubis).

Anterior View

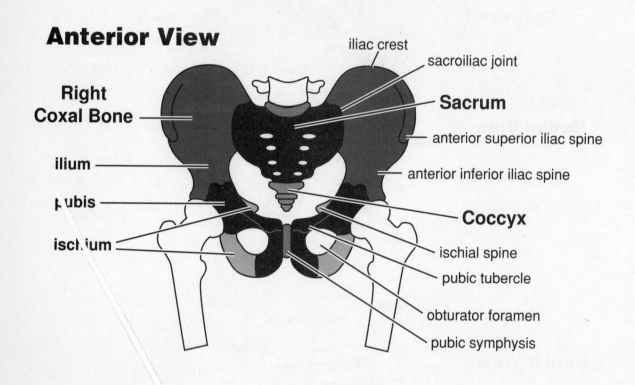

iliac crest
sacroiliac joint
Right Coxal Bone
Sacrum
ilium
anterior superior iliac spine
pubis
anterior inferior iliac spine
ischium
Coccyx
ischial spine
pubic tubercle
obturator foramen
pubic symphysis

Posterior View

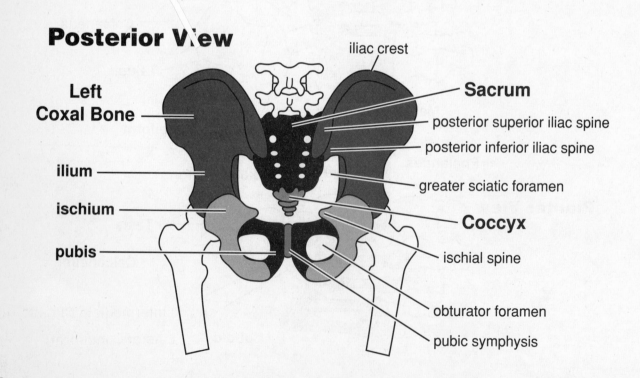

iliac crest
Left Coxal Bone
Sacrum
ilium
posterior superior iliac spine
posterior inferior iliac spine
ischium
greater sciatic foramen
pubis
Coccyx
ischial spine
obturator foramen
pubic symphysis

COXAL BONE and SACRUM

Right Coxal Bone
(lateral view)

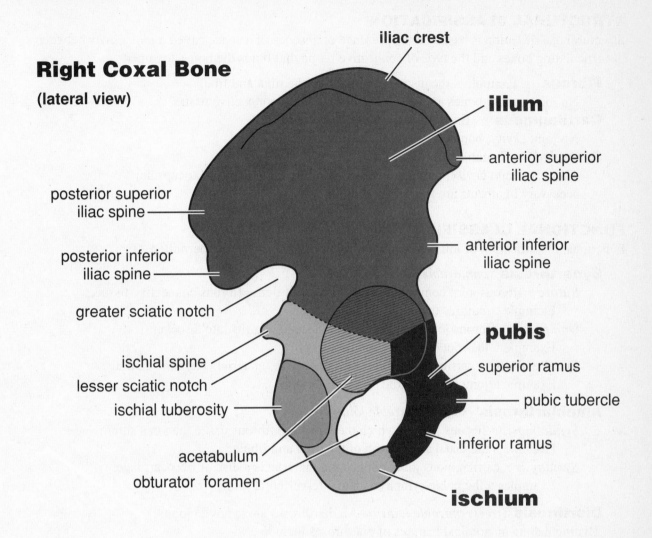

iliac crest

ilium

anterior superior
iliac spine

posterior superior
iliac spine

anterior inferior
iliac spine

posterior inferior
iliac spine

greater sciatic notch

pubis

ischial spine

superior ramus

lesser sciatic notch

pubic tubercle

ischial tuberosity

inferior ramus

acetabulum

obturator foramen

ischium

Sacrum and Coccyx

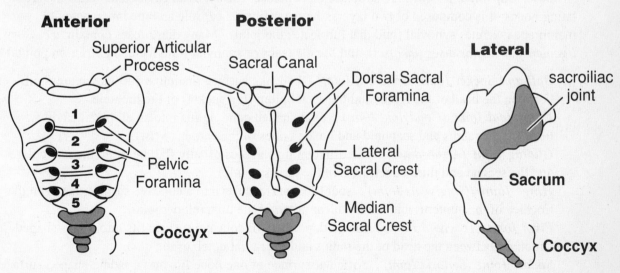

Anterior

Posterior

Superior Articular
Process

Sacral Canal

Dorsal Sacral
Foramina

Lateral

sacroiliac
joint

1
2
3
4
5

Pelvic
Foramina

Lateral
Sacral Crest

Sacrum

Median
Sacral Crest

Coccyx

Coccyx

SKELETAL SYSTEM / Joints (Articulations)

STRUCTURAL CLASSIFICATION

Structural classification is based on the presence or absence of a space called a *joint cavity* between the articulating bones and the type of connective tissue that binds the bones together.

Fibrous   Example : the distal articulation of the tibia and fibula.
No joint cavity; bones are held together by fibrous connective tissue.
Cartilaginous   Example : the pubic symphysis.
No joint cavity; bones are held together by cartilage.
Synovial   Example : the knee joint.
There is a joint cavity; bones are united by a surrounding articular capsule; accessory ligaments are frequently present.

FUNCTIONAL CLASSIFICATION

Functional classification of joints is based on the degree of movement permitted.

Synarthrosis *(immovable joint)*
Suture : fibrous joint composed of a thin layer of dense fibrous connective tissue.
    Example : sutures that unite skull bones.
Gomphosis : fibrous joint in which a cone-shaped peg fits into a socket.
    Example : the root of a tooth and its socket.
Synchondrosis : cartilaginous joint; the connective material is hyaline cartilage.
    Example : joint between true ribs and sternum.

Amphiarthrosis *(slightly movable joint)*
Syndesmosis : fibrous joint in which there is more fibrous tissue than in a suture.
    Example : the distal articulation of the tibia and fibula.
Symphysis : cartilaginous joint; connecting material is a disc of fibrocartilage.
    Examples : the pubic symphysis and intervertebral joints.

Diarthrosis *(freely movable joint)*   All diarthroses are synovial joints.
 Distinguishing anatomical features of diarthroses include :
Synovial Cavity (Joint Cavity) : a cavity that separates the articulating bones.
Articular Cartilage : a hyaline type cartilage that covers the bone surfaces at the joint.
Articular Capsule : a sleevelike structure that encloses the synovial cavity and unites the articulating bones; it is composed of two layers, the outer fibrous capsule and the inner synovial membrane (secretes synovial fluid that lubricates the joint). Many diarthroses contain *accessory ligaments*, *articular discs (menisci)*, and *bursae* (sacs of synovial fluid located at friction points).

Ball-and-Socket Joint (*Spheroid Joint*) : ball-like surface fits into a cup-like depression; between the head of the femur and the acetabulum (hip socket) of the hipbone.
Ellipsoidal Joint (*Condyloid Joint*) : oval-shaped condyle fits into an elliptical cavity; between the radius and scaphoid and lunate bones of the carpus (wrist).
Gliding Joint (*Arthrodial Joint*) : articulating surfaces usually flat; between the navicular and the second and third cuneiforms of the tarsus (ankle).
Hinge Joint (*Ginglymus Joint*) : spool-like surface fits into a concave surface; between the trochlea of the humerus and the trochlear notch of the ulna (elbow joint).
Pivot Joint (*Trochoid Joint*) : rounded, pointed, or concave surface fits into a ring-shaped structure; between the head of the radius and the radial notch of the ulna.
Saddle Joint (*Sellaris Joint*) : Articular surface of one bone fits into a saddle-shaped surface of another; between the trapezium of the carpus (wrist) and the metacarpal of the thumb.

JOINTS (Articulations)

Synarthrosis
(immovable joint)

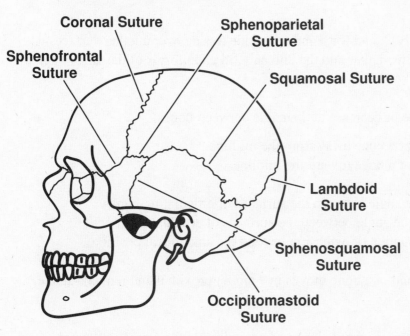

Coronal Suture

Sphenoparietal Suture

Sphenofrontal Suture

Squamosal Suture

Lambdoid Suture

Sphenosquamosal Suture

Occipitomastoid Suture

Amphiarthrosis
(slightly movable joint)

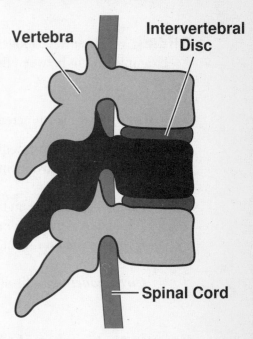

Vertebra

Intervertebral Disc

Spinal Cord

Diarthrosis (freely movable joint)

Synovial Joint
(generalized)

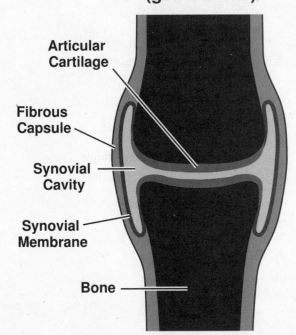

Articular Cartilage

Fibrous Capsule

Synovial Cavity

Synovial Membrane

Bone

Knee Joint

Patellofemoral : Synovial Gliding Type
Lateral Tibiofemoral : Synovial Hinge Type
Medial Tibiofemoral : Synovial Hinge Type

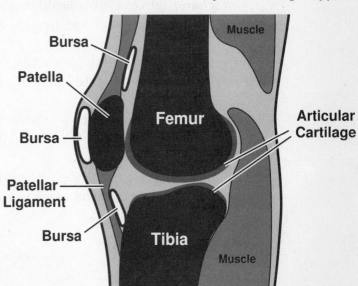

Bursa

Patella

Bursa

Patellar Ligament

Bursa

Muscle

Femur

Articular Cartilage

Tibia

Muscle

SKELETAL SYSTEM / Movements at Synovial Joints

Movements at synovial joints are produced by the actions of skeletal muscles. When the muscle contracts, the movable end (insertion) is pulled toward the fixed end (origin), and a movement occurs at the joint.

Gliding   One surface moves back and forth and from side to side over another surface.
   Example : joint between the navicular and the 2nd and 3rd cuneiforms of the tarsus (ankle).

Angular   There is an increase or decrease at the angle between bones.

   Abduction : movement of a bone away from the midline.
   Adduction : movement of a bone toward the midline.

   Flexion : decrease in the angle between the surfaces of articulating bones.
   Extension : increase in the angle between the surfaces of articulating bones.
   Hyperextension : continuation of extension beyond the anatomical position.

   Circumduction : distal end of a bone moves in a circle; proximal end remains stable.

Rotation   Movement of a bone around its longitudinal axis; may be medial or lateral.
   Example : rotational movement at ball-and-socket joints (shoulder and hip joints).

Special

   elevation : movement of a body part upward.
   depression : movement of a body part downward.

   protraction : movement of mandible or shoulder forward, parallel to the ground.
   retraction : movement of a protracted part backward, parallel to the ground.

   eversion : movement of the sole of the foot outward.
   inversion : movement of the sole of the foot inward.

   dorsiflexion : bending of the foot in the direction of the upper surface (dorsum).
   plantar flexion : bending of the foot in the direction of the sole (plantar surface).

   supination : turning the hand so the palm is upward.
   pronation : turning the hand so the palm is downward.

MOVEMENTS AT SYNOVIAL JOINTS

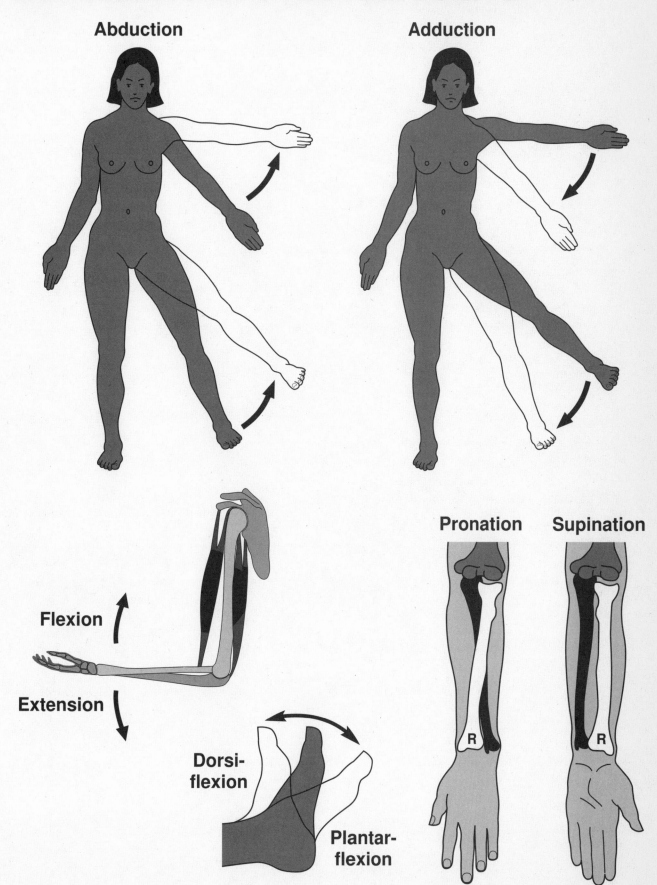

Abduction

Adduction

Flexion

Extension

Dorsi-flexion

Plantar-flexion

Pronation

Supination

2 Muscle Physiology

MUSCLE PHYSIOLOGY / Muscle Tissues

TYPES OF MUSCLE TISSUES

(1) Skeletal Muscle

Location : attached to bones (some attached to skin, deep fascia, or other muscles).
Microscopic Appearance : striated; many nuclei in each fiber (cell); unbranched fibers.
Fiber Diameter : 10 to 100 micrometers.
Fiber Length : 100 micrometers to 30 centimeters (about 1 foot).
Nervous Control : voluntary (conscious) control by somatic nervous system.
Regeneration : limited capacity; cells cannot divide.
Functions :
>   moves parts of the skeleton (walking, running, nodding the head, manipulating objects);
>   postural muscles maintain the body in stable positions;
>   the diaphragm regulates breathing by changing intrathoracic volume.

(2) Cardiac Muscle

Location : heart.
Microscopic Appearance : striated; single nucleus; branched fibers with intercalated discs.
Fiber Diameter : 14 micrometers.
Fiber Length : 50 to 100 micrometers.
Nervous Control : involuntary (unconscious) control by autonomic nervous system.
Hormonal Control : epinephrine & norepinephrine increase rate and strength of contractions.
Regeneration : none.
Function : propels blood through the blood vessels.

(3) Smooth Muscle

Location : walls of hollow organs; blood vessels; iris and ciliary muscles; arrector pili (hair).
Microscopic Appearance : no striations; single nucleus; spindle-shaped fibers.
Fiber Diameter : 3 to 8 micrometers.
Fiber Length : 30 micrometers to 200 micrometers.
Nervous Control : involuntary (unconscious) control by the autonomic nervous system.
Regeneration : more than other muscle tissues; much less than epithelial tissues.
Functions :
>   mixes and propels luminal contents through the digestive tract;
>   regulates the flow of blood through blood vessels by changing diameter of the lumen;
>   contraction of urinary bladder, gallbladder, and spleen expels urine, bile, and blood;
>   sphincter muscles control tube openings (anal sphincter, precapillary sphincters, etc.);
>   iris muscles control pupil diameter; ciliary muscles control lens shape;
>   contraction of arrector pili muscles causes hairs to stand up (goose pimples).

CHARACTERISTICS OF MUSCLE TISSUES

(1) Excitability (Irritability) : ability to generate action potentials in response to stimuli.
(2) Contractility : ability to contract and generate a force.
(3) Extensibility : ability to be stretched when pulled.
(4) Elasticity : ability to return to original length after contraction or extension.

MUSCLE TISSUE

Skeletal Muscle

Skeletal Muscle Cell (longitudinal section)

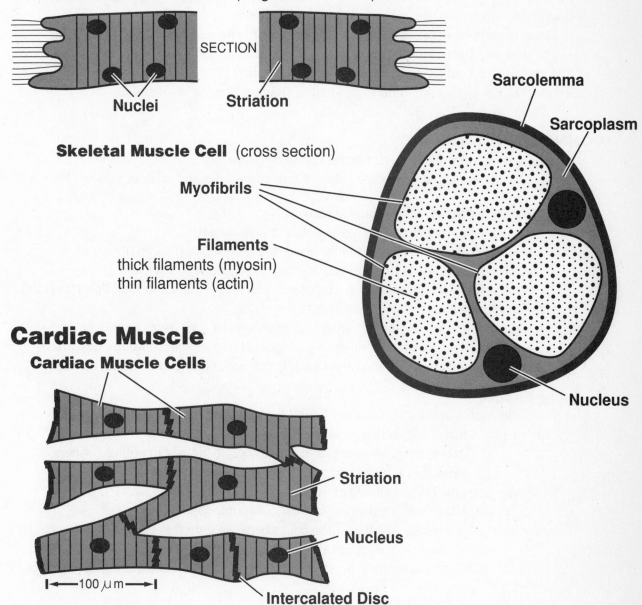

SECTION

Nuclei

Striation

Skeletal Muscle Cell (cross section)

Myofibrils

Filaments
thick filaments (myosin)
thin filaments (actin)

Sarcolemma

Sarcoplasm

Nucleus

Cardiac Muscle

Cardiac Muscle Cells

Striation

Nucleus

|←——100 μm——→|

Intercalated Disc

Smooth Muscle

Smooth Muscle Cells

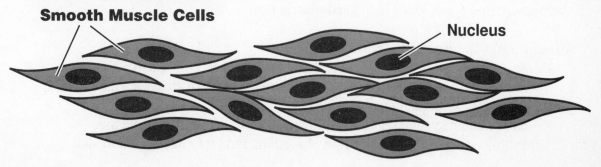

Nucleus

MUSCLE PHYSIOLOGY / Skeletal Muscle Anatomy

STRUCTURES

Fascicle (fasciculus) : a bundle of 10 to 100 skeletal muscle fibers.
Muscle Fiber : a muscle cell; a tube-shaped structure containing myofibrils.
Myofibrils : tube-shaped structures inside muscle fibers.
> The arrangement of protein filaments inside myofibrils causes dark and light bands.
> The dark and light bands on myofibrils give skeletal muscles their striated appearance;
> dark bands are called *A Bands*; light bands are called *I Bands*.

Sarcomeres : the repeating units composed of filaments inside myofibrils.

CONNECTIVE TISSUE

Fascia : a sheet or broad band of fibrous connective tissue.
> Fascia is located beneath the skin or around muscles and other organs.

Superficial Fascia   (Subcutaneous Layer)

composition : loose connective tissue (adipose and areolar).
> Adipose tissue consists of reticular fibers and adipocytes (fat cells);
> areolar tissue consists of fibers (collagen, elastic, and reticular) and
> many cell types (fibroblasts, macrophages, mast cells, adipocytes, and plasma cells).

location : immediately deep to the skin.
functions : fat and water storage; insulation (protects the body from heat loss);
> mechanical protection (cushions internal organs against traumatic blows);
> stabilizes nerves and blood vessels that are entering and exiting the muscles.

Deep Fascia

The components of muscle tissue are bound together by deep fascia.
composition : dense connective tissue (irregular).
> Dense, irregular connective tissue consists mostly of collagen fibers;
> some elastic fibers and a few fibroblasts are also present.

location : lines the body wall and extremities;
> surrounds whole muscles, fascicles, and individual fibers.

functions : holds muscles together; fills spaces between muscles;
> allows the free movement of muscles;
> stabilizes nerves and blood vessels that are entering and exiting the muscles.

Extensions of Deep Fascia

Epimysium : surrounds the entire muscle.
Perimysium : surrounds fascicles.
Endomysium : surrounds individual muscle fibers.

Attachments to Other Structures

Tendon : a cord of dense connective tissue that attaches a muscle to a bone.
Aponeurosis : a broad sheet of dense connective tissue that attaches a muscle to
> a bone, the skin, or another muscle.

Tendon Sheath (Synovial Sheath) : a tube of fibrous connective tissue that
> encloses a tendon; contains a film of synovial fluid (permits tendon to slide).

SKELETAL MUSCLE ANATOMY

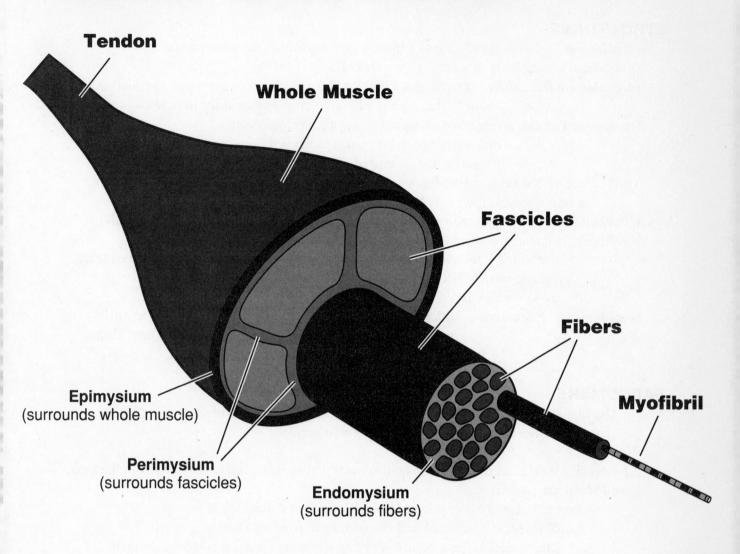

Tendon

Whole Muscle

Fascicles

Fibers

Myofibril

Epimysium
(surrounds whole muscle)

Perimysium
(surrounds fascicles)

Endomysium
(surrounds fibers)

Sarcomeres and Myofibril Bands

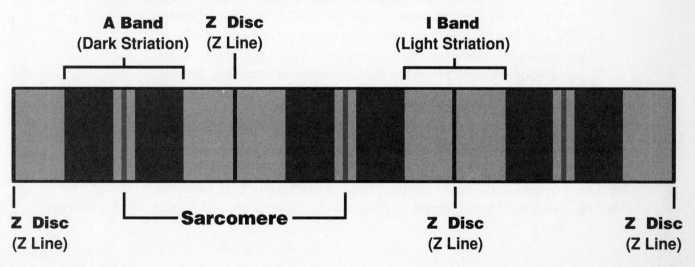

A Band
(Dark Striation)

Z Disc
(Z Line)

I Band
(Light Striation)

Z Disc
(Z Line)

Sarcomere

Z Disc
(Z Line)

Z Disc
(Z Line)

MUSCLE PHYSIOLOGY / Skeletal Muscle Fiber

STRUCTURES

Sarcolemma : plasma membrane of a muscle cell; surrounds the sarcoplasm.

Sarcoplasm : cytoplasm of a skeletal muscle cell.

Sarcoplasmic Reticulum : a network of smooth endoplasmic reticulum that surrounds each myofibril; stores calcium that is essential for muscle contraction.

Transverse Tubules : fingerlike invaginations of the sarcolemma; they carry action potentials to the sarcoplasmic reticulum, triggering the release of calcium ions.

Nuclei : each fiber has many nuclei located at the periphery of the cell in the sarcoplasm; a single muscle fiber may have as many as 100 to 1,000 nuclei.

Mitochondria : a fiber has many mitochondria that lie in rows throughout the sarcoplasm.

Myofibrils : tubular structures that extend lengthwise within the muscle fiber; alternating light and dark bands give myofibrils a striped (striated) appearance; a single muscle fiber may contain between 300 and 3000 myofibrils, each of which is 1 to 2 micrometers in diameter.

Myofilaments (Filaments) : within each myofibril are threads of protein called myofilaments or filaments; there are 3 types : thin, thick, and elastic.

SARCOMERE

The filaments inside a myofibril do not extend the entire length of the muscle fiber; they are arranged in repeating units called sarcomeres.

Z Discs (Z Lines) : narrow, plate-shaped regions that separate one sarcomere from the next.

Thin Filaments (Actin) :
A thin filament consists primarily of bean-shaped molecules of *actin*;
a thin filament is a strand of actin molecules twisted into a helix;
each actin molecule has a myosin-binding site where a cross-bridge can attach;
thin filaments also contain regulatory proteins : tropomyosin and troponin;
thin filaments extend from anchoring points within Z discs.

Thick Filaments (Myosin) :
A thick filament consists of about 200 molecules of the protein *myosin*;
each myosin molecule is shaped like two golf clubs twisted together;
the golf club handles point toward the center of the sarcomere;
cross-bridges (golf club heads) point toward adjacent thin filaments;
cross-bridges project from the thick filaments in a spiraling fashion.

Elastic Filaments (Titin) :
Elastic filaments consist of the protein *titin*; they anchor thick filaments to Z discs.

A Band (dark band) : orderly parallel arrangement of thick filaments; middle of sarcomere.

I Band (light band) : includes segments of two adjacent sarcomeres; contains portions of thin filaments that do not overlap thick filaments.

H Zone : a band in the center of the A band; contains only thick filaments.

M Line : divides H zone; formed by protein molecules that connect adjacent thick filaments.

SKELETAL MUSCLE FIBER

Portion of a Muscle Fiber

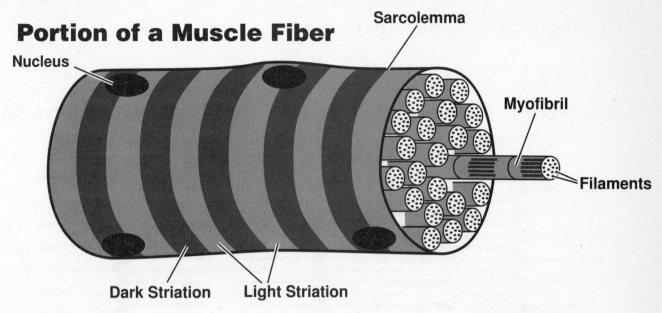

Sarcolemma

Nucleus

Myofibril

Filaments

Dark Striation   Light Striation

Myofibril

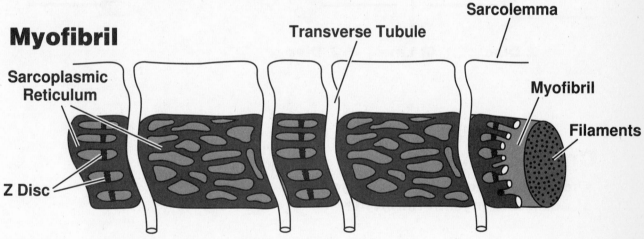

Sarcolemma

Transverse Tubule

Sarcoplasmic Reticulum

Myofibril

Filaments

Z Disc

Sarcomeres

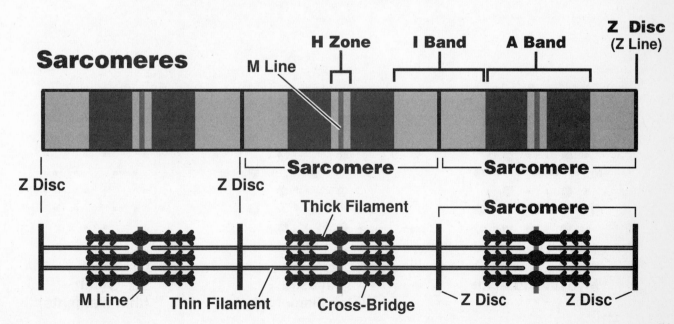

H Zone   I Band   A Band   Z Disc (Z Line)

M Line

Z Disc   Z Disc   Sarcomere   Sarcomere

Sarcomere

Thick Filament

M Line   Thin Filament   Cross-Bridge   Z Disc   Z Disc

MYOFILAMENTS

Longitudinal Section

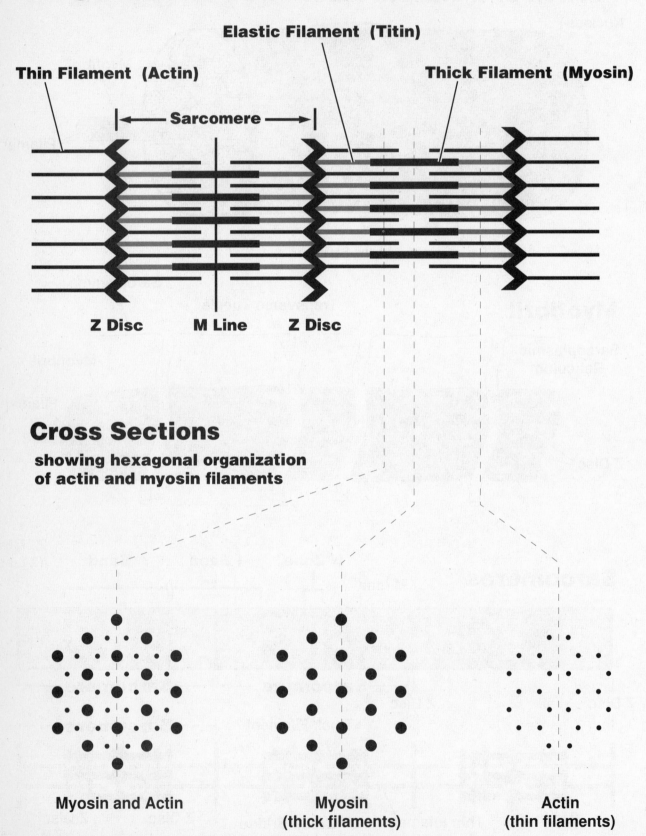

Elastic Filament (Titin)

Thin Filament (Actin)

Thick Filament (Myosin)

Sarcomere

Z Disc   M Line   Z Disc

Cross Sections
showing hexagonal organization
of actin and myosin filaments

Myosin and Actin

Myosin
(thick filaments)

Actin
(thin filaments)

ARRANGEMENT OF FASCICULI

Fasciculi : bundles of skeletal muscle fibers.
Skeletal muscle fibers are arranged in a parallel fashion within each bundle, but the arrangement of the fasciculi with respect to the tendons may take several characteristic patterns : parallel, circular, fusiform, and pennate.

Parallel

fasciculi are parallel with longitudinal axis of muscle and terminate at either end in flat tendons

Circular

fasciculi are arranged in a circular pattern

Fusiform

fasciculi nearly parallel with longitudinal axis and muscle tapers toward tendons

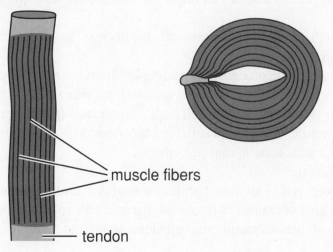

muscle fibers

tendon

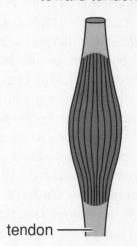

tendon

Unipennate

fasciculi are arranged on only one side of tendon

Bipennate

fasciculi are arranged on both sides of centrally positioned tendon

Multipennate

fasciculi attach obliquely from many directions to several tendons

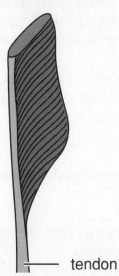

tendon

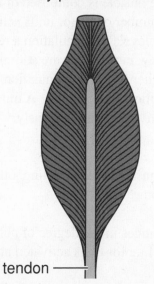

tendon

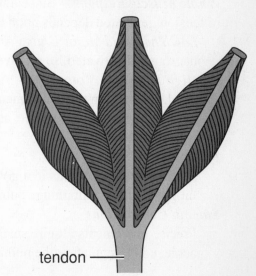

tendon

Motor Unit : a motor neuron and the muscle fibers it stimulates;
         a motor unit may contain as few as 2 muscle fibers and as many as 2000.

Fine Muscle Control : relatively few muscle fibers stimulated by one motor neuron;
     a single motor neuron may supply just 10 muscle fibers in the eye.
     As a result, there is finer control over the amount of tension in the whole muscle.

Coarse Muscle Control : many muscle fibers stimulated by one motor neuron;
     a single motor neuron may supply 2000 muscle fibers in the gastrocneumius.
     As a result, there is less control over the amount of tension in the whole muscle.

Structures

Motor Neuron : any neuron (nerve cell) that carries impulses from the brain or spinal cord to an effector (muscle or gland).

Cell Body : the part of a neuron that immediately surrounds the nucleus; motor neuron cell bodies are located in the brain or spinal cord.

Dendrites : highly branched cytoplasmic processes extending out from the neuron cell body; they increase the surface area of the neuron, so it can receive input from many other neurons.

Axon : a long process extending out from a neuron cell body. It conducts impulses away from the cell body to other cells (muscle cells, gland cells, or other neurons). Axons branch just before they come into contact with skeletal muscle cells.

Axon Terminal : the end of an axon.

Synaptic End Bulb : the expanded end of an axon terminal; contains neurotransmitter.

Neuromuscular Junction : the area of contact between the synaptic end bulb of a motor neuron and a portion of the muscle cell plasma membrane called the motor end plate.

Recruitment of Motor Units

Recruitment : the process of increasing the number of active motor units.

Single Motor Unit Response Since all muscle fibers in a motor unit are stimulated by the same motor neuron, they all contract simultaneously.

Whole Muscle Response Since a whole muscle is composed of many motor units, the amount of tension generated depends upon the number of motor units activated at the same time.

    Low Threshold Motor Units At low levels of stimulation a relatively small number of motor neurons are activated, so relatively few motor units are stimulated.

    High Threshold Motor Units At higher levels of stimulation more motor neurons are stimulated, resulting in the recruitment of more motor units. A muscle is contracting at maximal intensity when all motor units are activated simultaneously.

Prevention of Fatigue
While some motor units in a given muscle are contracting, others are relaxed; this prevents fatigue, while maintaining contraction.

Smooth Contractions
Precise movements require small changes in the degree of contraction of a muscle; this is achieved by regulating the number of motor units activated at a given time.

MOTOR UNITS

Motor Unit : the number of muscle fibers innervated by a single motor neuron

Fine Muscle Control

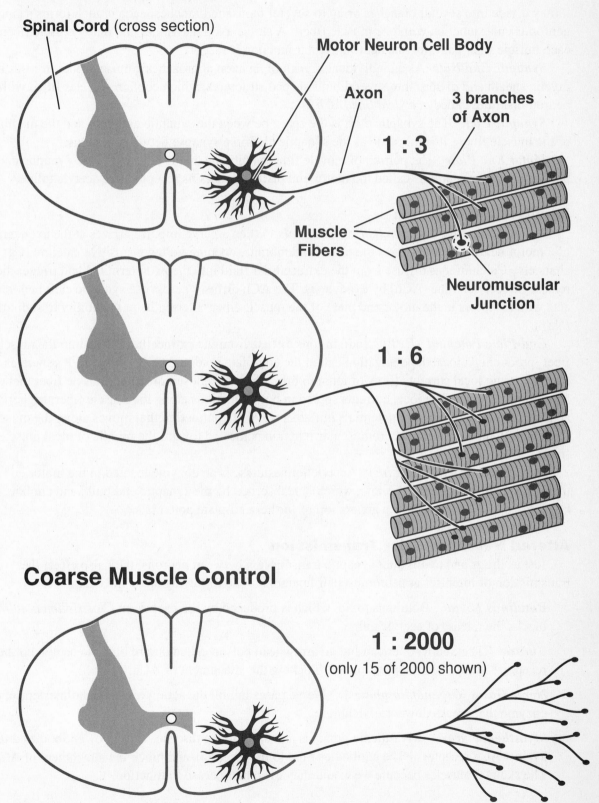

Spinal Cord (cross section)

Motor Neuron Cell Body

Axon

3 branches of Axon

1 : 3

Muscle Fibers

Neuromuscular Junction

1 : 6

Coarse Muscle Control

1 : 2000
(only 15 of 2000 shown)

MUSCLE PHYSIOLOGY / Neuromuscular Junction

Structures

The contact point between a motor neuron and a skeletal muscle fiber is called a *neuromuscular junction* or *myoneural junction*. As the axon of a single motor neuron approaches a skeletal muscle, it may divide into several branches or up to several thousand branches; each branch forms a single neuromuscular junction with one muscle fiber. A single axon may innervate many muscle fibers, but each muscle fiber has only one neuromuscular junction.

Synaptic End Bulb As an individual branch of an axon approaches a muscle fiber, it loses its myelin sheath and divides into several bulb-shaped structures, which contain vesicles filled with a neurotransmitter called *acetylcholine (ACh)*.

Synaptic Cleft The synaptic cleft is the space between the synaptic end bulb and the membrane of the muscle fiber. It is the same as the synaptic cleft in a synapse between neurons.

Motor End Plate The portion of muscle fiber membrane that is adjacent to the synaptic end bulb of its motor neuron is called the motor end plate. It contains receptors for acetylcholine.

Impulse Transmission

Voltage-Sensitive Calcium (Ca^{+2}) Channels When a nerve impulse arrives at the axon terminal of a motor neuron, it depolarizes the plasma membrane, opening voltage-sensitive calcium (Ca^{+2}) channels. Calcium ions diffuse from the extracellular fluid into the axon terminal and trigger the release of acetylcholine (ACh) by exocytosis. The ACh diffuses across the synaptic cleft and inter- acts with receptors in the motor end plate of the muscle fiber, altering its permeability to sodium ions (Na^+).

End-Plate Potential (EPP) Sodium ions diffuse from the extracellular fluid into the muscle fiber, producing a local depolarization called the end-plate potential (EPP). One EPP generates strong enough local currents (flow of ions) to bring the adjacent sarcolemma (muscle fiber mem- brane) to threshold. The local currents spread in both directions along the muscle fiber, triggering action potentials. The action potentials initiate a wave of contraction that moves along the muscle fiber in both directions (the neuromuscular junction is located toward the middle of most muscle fibers).

Acetylcholinesterase (AChE) Acetylcholinesterase is an enzyme located in the motor end plate that splits acetylcholine into acetate, which is reabsorbed by the synaptic end bulb, and choline. The breakdown of ACh prevents the generation of further end-plate potentials.

Altered Neuromuscular Transmission

Just as drugs and toxins alter synaptic transmission between neurons, they also effect the transmission of impulses at neuromuscular junctions. The following are some examples:

Botulinus Toxin Botulinus toxin, which is produced by the bacterium *Clostridium botulinum*, blocks the release of acetylcholine.

Curare The South American Indian arrowhead poison called curare binds to acetylcholine receptor sites in the motor end plates, blocking the attachment of ACh.

Nerve Gases (Organophosphates) Nerve gases inhibit the action of acetylcholinesterase, the enzyme that breaks down acetylcholine.

Myasthenia Gravis Myasthenia gravis is an autoimmune disorder caused by antibodies directed against ACh receptors. The antibodies bind to the receptors and block the attachment of ACh. The skeletal muscles become weak and may eventually cease to function.

NEUROMUSCULAR JUNCTION

Structures

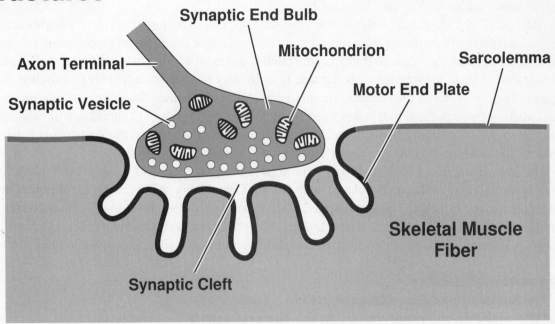

Impulse Transmission

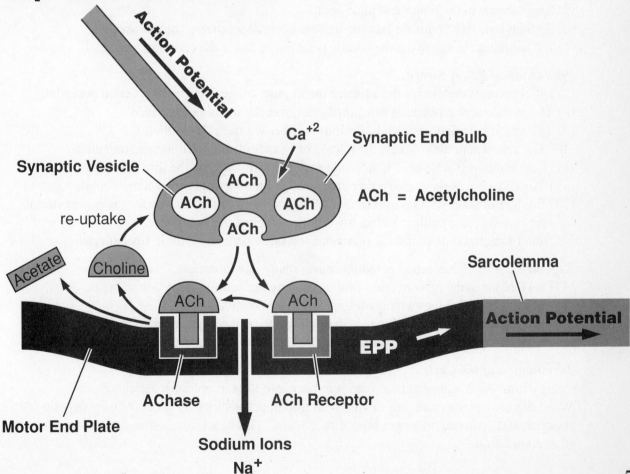

MUSCLE PHYSIOLOGY / Excitation-Contraction Coupling

Excitation-contraction coupling refers to the sequence of events by which an end-plate potential leads to cross-bridge activity in a muscle fiber.

After acetylcholine is released from the synaptic end bulb of a motor neuron, it diffuses across the synaptic cleft and combines with receptors on the surface of the adjacent sarcolemma (the motor end plate). The end-plate potential that is generated triggers an action potential that travels along the sarcolemma and into the interior of the muscle fiber by way of transverse tubules (T-tubules). When the action potential reaches the sarcoplasmic reticulum it causes the release of calcium ions, which diffuse through the sarcoplasm to nearby molecules of troponin. Combination of calcium with troponin triggers the sliding filament mechanism; thin filaments slide past thick filaments toward the center of the sarcomeres, the sarcomeres shorten, and the muscle fiber shortens.

The immediate, direct source of energy for muscle contraction is ATP (adenosine *tri*phosphate). An enzyme (ATPase) splits a phosphate group from ATP, forming ADP (adenosine *di*phosphate) and P (phosphate group); the energy released when P is split from a molecule of ATP, activates (energizes) myosin cross-bridges. The activated cross-bridges attach to actin molecules (thin filaments) and rotate, causing the thin filaments to slide past the thick filaments (muscle fiber shortens).

Sequence of Events

Production of End-Plate Potential (EPP)
(1) A motor neuron action potential travels along the axon to the synaptic end bulb.
(2) Acetylcholine is released from the synaptic end bulb.
(3) Acetylcholine diffuses across the synaptic cleft to the motor end plate.
(4) Acetylcholine binds to receptor sites on the motor end plate.
(5) Ion channels in the motor end plate open.
(6) Sodium ions (Na^+) diffuse into the muscle fiber, depolarizing the membrane;
    the resulting change in the membrane potential is called the end-plate potential.

Production of Power Stroke
(7) Local currents depolarize the adjacent membrane, triggering a muscle action potential.
(8) The muscle action potential self-propagates over the membrane surface.
(9) The action potential passes into the muscle fiber via transverse tubules.
(10) The action potential triggers the release of calcium from sarcoplasmic reticulum.
(11) Calcium ions (Ca^{+2}) bind to troponin molecules (located on the thin filaments).
(12) Tropomyosin moves, uncovering cross-bridge binding sites on actin molecules.
(13) Binding of actin with myosin causes ATP to split, releasing energy for the power stroke;
    the myosin cross-bridges swivel toward the center of the sarcomere;
    thin filaments slide past thick filaments; the sarcomere and muscle fiber shorten.

Repeated Cycle   (One action potential causes many power strokes.)
(14) ATP binds to the myosin cross-bridge, causing the actin-myosin link to break.
(15) Myosin cross-bridge swivels back into original position.
(16) Myosin cross-bridge combines with binding site on the adjacent actin molecule.
(17) ATP splits, releasing energy that activates the cross-bridge, causing a power stroke.

Termination of the Cycle
Energy from ATP is used to pump calcium back into the sarcoplasmic reticulum.
When there are no calcium ions to bind with troponin, the blocking action of tropomyosin is restored, and actin can no longer bind with myosin. This blocks the sequence and stops muscle fiber contraction.

EXCITATION - CONTRACTION COUPLING

Muscle Fiber with Transverse Tubule

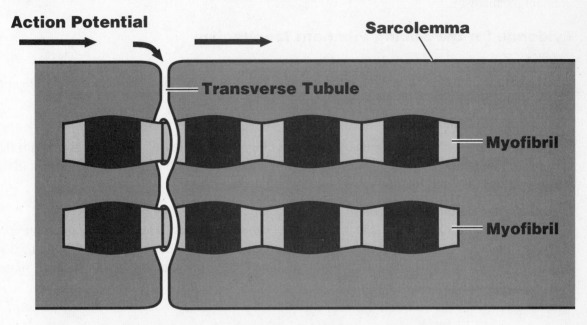

Calcium is released by sarcoplasmic reticulum.

Molecular Events

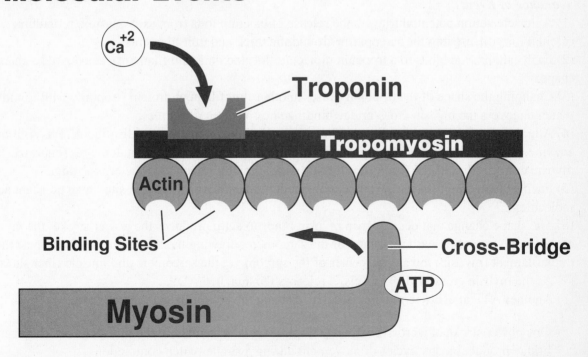

MUSCLE PHYSIOLOGY / Sliding Filament Mechanism

During muscle shortening, neither the thick filaments nor the thin filaments change in length. Instead, the thin filaments slide past the thick filaments. This action is referred to as the sliding filament mechanism.

Evidence for the Sliding Filament Mechanism

A Band (dark band)

When a skeletal muscle fiber shortens, the width of the A band remains constant. This width corresponds to the length of the thick filaments, which do not shorten.

I Band (light band)

When a skeletal muscle fiber shortens, the width of the I band decreases. The I band contains the portions of thin filaments that do not overlap thick filaments. As more and more of the thin filament length overlaps the thick filaments, the I band width decreases.

H Zone (center portion of dark band)

When a skeletal muscle shortens, the width of the H zone decreases. The H zone is the center portion of the dark band where only thick filaments are present (there is no overlapping of thin filaments). As thin filaments slide toward the center of a sarcomere they increasingly overlap the thick filaments, decreasing the width of the H zone.

Power Stroke

The power stroke is the rotational movement of a myosin cross-bridge. Because the cross-bridge is attached to an adjacent actin molecule, the power stroke moves the thin filament. The immediate, direct source of energy for the power stroke is ATP (adenosine triphosphate).

Sequence of Events :

(1) A muscle action potential triggers the release of calcium ions from sarcoplasmic reticulum; calcium ions diffuse into the sarcoplasm around the thick and thin filaments.

(2) Each calcium ion binds to a troponin molecule (located on a thin filament), causing it to change shape.

(3) Changing the shape of the troponin causes another thin filament protein (tropomyosin) to move, which uncovers the myosin cross-bridge binding sites on actin molecules.

(4) A molecule of ATP attaches to a binding site on each myosin cross-bridge; the ATP is split by an enzyme (ATPase) located on the myosin cross-bridge, releasing energy that activates (energizes) the myosin cross-bridge. ADP (adenosine diphosphate) remains attached to the cross-bridge.

(5) Each activated myosin cross-bridge spontaneously attaches to a binding site on an adjacent actin molecule.

(6) The shape change that occurs when myosin binds to actin produces the power stroke; the myosin cross-bridge swivels toward the center of the sarcomere, drawing the thin filament (actin) past the thick filament (myosin) toward the center of the sarcomere; the sarcomere and muscle fiber shorten.

(7) As the myosin cross-bridge swivels, it releases the attached ADP.

(8) Another ATP attaches to the cross-bridge, causing the actin-myosin link to break.

One power stroke of a cross-bridge results in a small movement of the thin filament. Each cross-bridge produces many cycles of movement during a single twitch contraction.

SLIDING FILAMENT MECHANISM

Relaxed Muscle Fiber

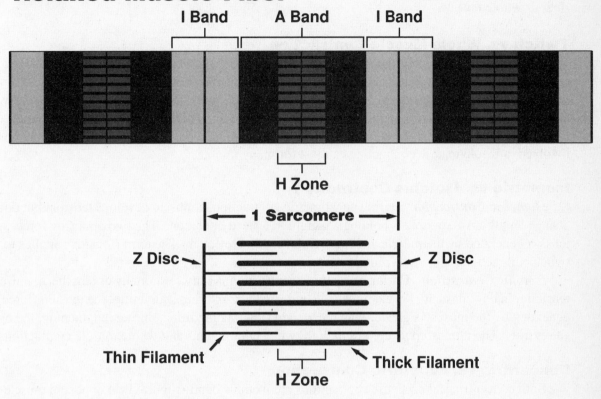

I Band | A Band | I Band

H Zone

1 Sarcomere

Z Disc

Z Disc

Thin Filament

Thick Filament

H Zone

Contracted Muscle Fiber

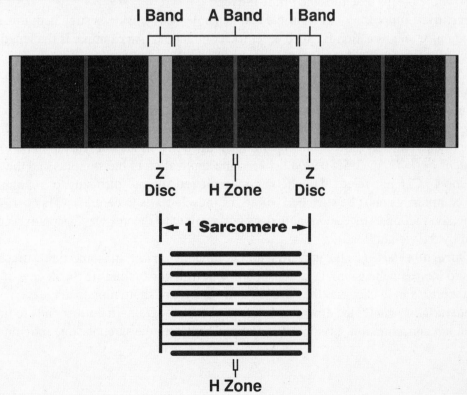

I Band | A Band | I Band

Z Disc | H Zone | Z Disc

1 Sarcomere

H Zone

MUSCLE PHYSIOLOGY / Muscle Contractions

Contraction : activation of the tension-generating process that results in a shortening of the muscle fiber or whole muscle.

Twitch vs. Whole Muscle Contraction

Twitch Contraction A twitch is a brief contraction of all the muscle fibers in a motor unit in response to a single motor neuron action potential. It lasts only a fraction of a second.

Whole Muscle Contraction A whole muscle contraction is the mechanical response of a muscle to many motor neuron action potentials that activate thousands of muscle fibers in many motor units. The duration of a whole muscle contraction depends upon the endurance of the fibers in the muscle involved; a whole muscle contraction can last for many minutes.

Isometric vs. Isotonic Contraction

Isometric Contraction During an isometric contraction, a muscle develops tension, but does not change length. An example is holding a weight in a steady position. The two opposing forces, the tension generated by the muscle and the weight (load), are equal. The term *isometric* applies to both twitch contractions and whole muscle contractions.

Isotonic Contraction During an isotonic contraction, a muscle shortens or lengthens; muscle tension remains constant. An example is lifting a weight. During shortening, the tension (force) generated by the muscle is greater than the weight (load) being lifted; during lengthening, the opposite is true. The term *isotonic* applies to both twitch contractions and whole muscle contractions.

Concentric vs. Eccentric Contraction

Both concentric and eccentric contractions are isotonic contractions. Two opposing muscles contracting at the same time with unequal forces results in a smooth movement.

Concentric Contraction The muscle undergoing the strongest contraction shortens; this muscle is called the *prime mover* and its action is called concentric contraction. During flexion of the arm at the elbow, the biceps brachii is the prime mover.

Eccentric Contraction The muscle that generates less tension is stretched; this muscle is called the *antagonist* and its action is called eccentric contraction. An example is the lengthening of the triceps brachii during flexion of the arm at the elbow.

Myogram

A myogram is a graph of a muscle contraction. The terminology and methods of recording contractions apply to both single muscle fibers and whole muscles. The line (trace) on a myogram may be divided into 3 phases : latent period, contraction period, and relaxation period.

Latent Period : the brief period between the application of the stimulus and the start of the contraction. Calcium released by the sarcoplasmic reticulum is diffusing to troponin molecules; also elastic components must be stretched before the muscle starts to contract. This period is longer in isotonic contractions, since shortening does not begin until the muscle tension exceeds the resistance offered by the weight (load).

Contraction Period : the upward tracing of the myogram. In isometric contractions it corresponds to increasing tension. In isotonic contractions it corresponds to the degree of muscle shortening; an increase in load causes the duration and velocity of shortening to decrease.

Relaxation Period : the downward tracing of the myogram. It corresponds to the decrease in calcium ion concentration, as calcium is pumped back into the sarcoplasmic reticulum.

ISOMETRIC VS. ISOTONIC CONTRACTION

Isometric Contraction
Muscle length remains constant, tension changes

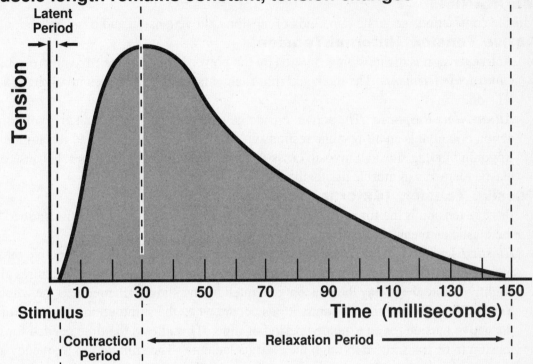

Latent Period

Tension

Stimulus

10   30   50   70   90   110   130   150

Time (milliseconds)

Contraction Period

Relaxation Period

Isotonic Contraction
Tension remains constant, muscle length changes

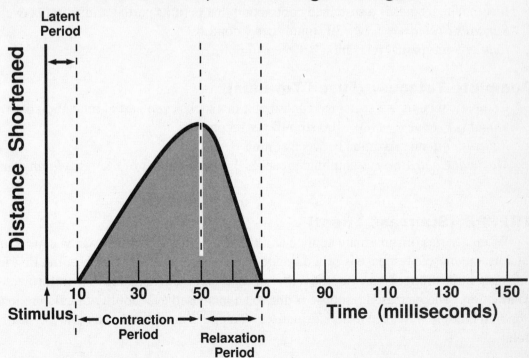

Latent Period

Distance Shortened

Stimulus

10   30   50   70   90   110   130   150

Time (milliseconds)

Contraction Period

Relaxation Period

MUSCLE PHYSIOLOGY / Summation and Treppe

Wave Summation   Wave summation is the increased strength of a contraction resulting from the application of a second stimulus before the muscle has completely relaxed.

MECHANISM
Muscle contractions generate two kinds of tension : active tension and passive tension.

Active Tension (Internal Tension)
Active tension is the tension generated by the power strokes of the myosin cross-bridges.
Contractile Elements The thick and thin filaments are the structures involved in active tension.
All-or-None Response The active tension generated in response to a single motor neuron action potential is an all-or-none response : enough calcium is released to saturate all of the troponin binding sites; all myosin cross-bridges are activated. However, because of the elastic elements in muscle tissue, the muscle does not fully contract.

Passive Tension (External Tension)
Passive tension is the force (pull) that results after the active tension is transmitted through the elastic elements of the muscle.
Elastic Elements Elastic elements include elastic filaments (titin), connective tissue coverings (epimysium, perimysium, and endomysium), and tendons. These structures all stretch slightly before they relay the tension generated by the sliding filaments to the whole muscle.
Summation Summation of contractions occurs when the stimulation frequency maintains the active tension for an extended period of time. This allows the time needed for the active tension to be transmitted through the elastic elements. The stimulation frequency at which summation occurs varies with the type of muscle fiber (slow-twitch or fast-twitch) involved. Summation results from the progressive build up of calcium ions in the sarcoplasm.

SUMMATION
Incomplete Tetanus   (Unfused Tetanus)
Incomplete tetanus is a sustained contraction that permits partial relaxation between stimuli.
Stimulus Frequency : 10 - 30 stimuli per second.
Relaxation : partial relaxation between stimuli.

Complete Tetanus   (Fused Tetanus)
Complete tetanus is a sustained contraction that lacks even partial relaxation between stimuli.
Stimulus Frequency : 80 - 100 stimuli per second.
Relaxation : no relaxation between stimuli.
Maximal Contraction :   amplitude reaches 3 to 4 times that of a single twitch contraction.

TREPPE (Staircase Effect)
When a series of stimuli are applied at a frequency just below that which causes incomplete tetanus, the muscle response is called treppe or the staircase effect. Each of the first few contractions is a little stronger than the last. After several contractions, there is a uniform tension per contraction. The increased response is due to an increased availablility of calcium ions for troponin binding, just as in tetanic responses. Treppe is not considered to be a type of summation.

SUMMATION

Incomplete Tetanus

10 stimuli per second
(1 stimulus per 100 milliseconds)

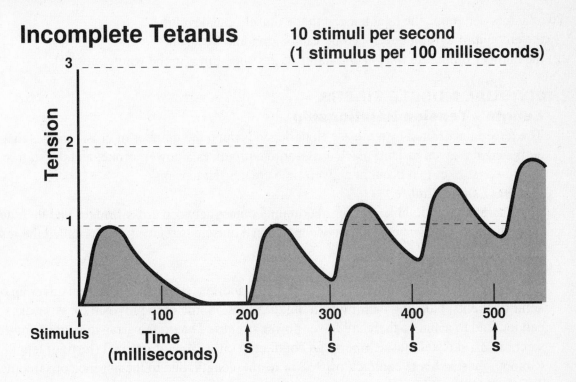

Complete Tetanus

100 stimuli per second
(10 stimuli per 100 milliseconds)

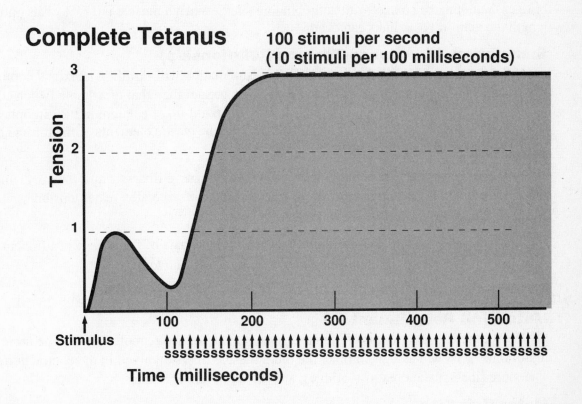

MUSCLE PHYSIOLOGY / Muscle Tension

Two factors determine the total tension that a muscle can develop:
(1) the amount of tension developed by each individual fiber; and
(2) the number of muscle fibers contracting at the same time (in the whole muscle).

INDIVIDUAL MUSCLE FIBERS
Length - Tension Relationship
The tension generated by a muscle fiber depends upon the number of cross-bridges that are activated at a given instant. Each cross-bridge produces a power stroke, and the sum of all the power strokes that occur at a given time creates the tension.

Optimal Length (100 %)

There is an optimal length for maximum contact between cross-bridges and their actin binding sites. Greatest contaction occurs when a muscle is relaxed; this is called the resting or optimal length.

70 % of Optimal Length

When a muscle is shortened, the sarcomeres shorten; some actin filaments overlap other actin filaments, blocking their myosin binding sites. Some of the myosin cross-bridges cannot bind to actin, so there are fewer power strokes. Due to the arrangement of end-to end sarcomeres, skeletal muscle fibers can contract to only 70 % of optimal (resting) length. (Smooth muscle fibers contract to 10 % of resting length, due to the absence of sarcomeres.)

175 % of Optimal Length

When a muscle is stretched, the sarcomeres are stretched, and some of the myosin cross-bridges cannot make contact with actin binding sites. When a muscle is 175% of its optimal length, no actin-myosin links can occur.

Summation (Frequency-Tension Relationship)
If the frequency of stimulation is sufficient, summation will result. A maximal contraction reaches an amplitude that is three to four times greater than that of a single twitch contraction. Greater tension is generated because of the build-up of calcium in the sarcoplasm; this allows time for tension to be transmitted through the elastic elements of the muscle fiber.

Fiber Type
Fast glycolytic fibers have larger diameters than oxidative fibers. Thus, they contain more thick and thin filaments, more cross-bridges, and can generate greater tension.

State of Fatigue
The extent of previous muscle activity determines the state of fatigue. If insufficient ATP is available, a muscle fiber cannot generate its maximum tension.

WHOLE MUSCLE   (the number of fibers contracting at the same time)
Motor Unit Recruitment
Recruitment is the process of increasing the number of active motor units. The more motor units activated at a given time, the more muscle fibers involved in the contraction. The more fibers, the more cross-bridges, and the more power strokes.

Motor Unit Size
The number of muscle fibers in a motor unit varies considerably. In finely controlled muscles the motor units are small; in coarsely controlled muscles the motor units are large.

LENGTH - TENSION RELATIONSHIP

Isometric Tetanus Tension
Variation in tension with muscle fiber length

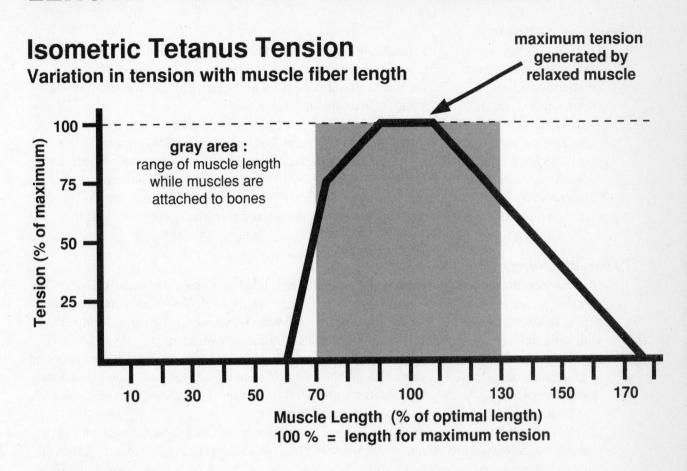

maximum tension
generated by
relaxed muscle

gray area :
range of muscle length
while muscles are
attached to bones

Tension (% of maximum)

100

75

50

25

10   30   50   70   100   130   150   170

Muscle Length (% of optimal length)
100 % = length for maximum tension

Sarcomere

70 % of optimal length
thin filaments overlap
blocking actin-myosin binding sites

100 % (the optimal length)
maximum contact between
myosin cross-bridges & actin

Z Disc

175 % of optimal length
no thin-thick filament overlap
no actin-myosin links can occur

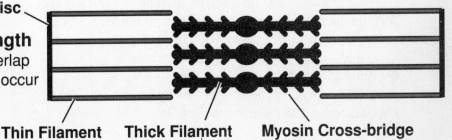

Thin Filament   **Thick Filament**   **Myosin Cross-bridge**

67

MUSCLE PHYSIOLOGY / Energy

Sources of Energy

Skeletal muscles increase ATP production on demand. There are 3 important sources of ATP :

(1) Phosphagen System : the use of creatine phosphate and stored ATP. A small amount of creatine phosphate and ATP is stored in muscle fibers for quick energy. This system provides enough energy for about 15 seconds of maximal muscle activity.

(2) Glycogen-Lactic Acid System : the conversion of glycogen or glucose into pyruvic acid or lactic acid by glycolysis. (When no oxygen is present, lactic acid is the final product.) This system yields 2 molecules of ATP from each glucose molecule and provides enough energy for about 30 seconds of maximal muscle activity. It is the major source of energy during a sprint.

(3) Aerobic System : the conversion of pyruvic acid into carbon dioxide, water, and ATP. It yields 36 molecules of ATP from each glucose molecule and provides energy for muscular activity lasting longer than 30 seconds. It is used during long distance running.

Uses of Energy

(1) Energizes the Myosin Cross-Bridge   The globular head of myosin (the cross-bridge) contains an enzyme (ATPase) that catalyzes the breakdown of ATP. The energy released from the breakdown of ATP energizes the myosin cross-bridge. When the myosin cross-bridge binds with actin, the energy is released, causing the rotation of the cross-bridge.

(2) Breaks the Myosin-Actin Link   During the power stroke, the myosin cross-bridge is bound to an adjacent actin molecule. This linkage must be broken at the end of each power stroke so that the myosin cross-bridge can bind to another actin molecule. The binding of a new molecule of ATP to myosin causes the myosin-actin link to break.

(3) Pumps Calcium into Sarcoplasmic Reticulum   As soon as calcium is released, an active transport system begins to pump the calcium back into the sarcoplasmic reticulum. There is an equilibrium between free calcium ions in the cytosol and calcium ions bound to troponin. Free calcium ions in the cytosol are constantly pumped into the sarcoplasmic reticulum.

Heat Production (Thermogenesis)   As much as 85% of the energy produced by cellular respiration during muscle contraction is released as heat.

Recovery Oxygen Consumption (Oxygen Debt)

The elevated oxygen intake after exercise (panting) is needed for the following uses :

(1) Conversion of lactic acid back into pyruvic acid.

(2) Re-establishment of glycogen stores in muscle cells and liver cells.

(3) Resynthesis of creatine phosphate and ATP stored in muscle cells.

(4) Replacement of oxygen removed from myoglobin (oxygen-storing protein in muscle cells).

(5) ATP production for metabolic reactions (increased rate due to increased body temperature).

(6) ATP production for the continuing elevated activity of cardiac and skeletal muscles.

(7) ATP production needed for an increased rate of tissue repair.

Fatigue

Inability of a muscle to maintain its strength of contraction or tension results from insufficient ATP production. Factors that contribute to fatigue include :

(1) Insufficient oxygen delivered to muscle cells.

(2) Depletion of glycogen stored in muscle cells.

(3) Buildup of lactic acid in body fluids.

(4) Insufficient acetylcholine released by synaptic end bulbs of motor neurons.

(5) Unexplained mechanisms in the brain.

ENERGY SOURCES

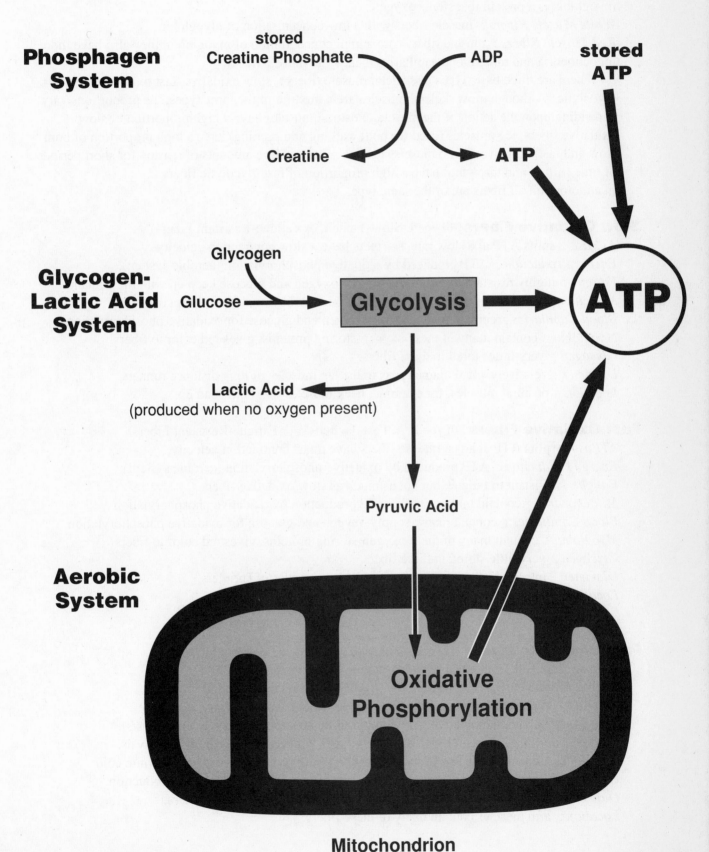

Phosphagen System

stored Creatine Phosphate → Creatine

ADP → ATP

stored ATP

Glycogen-Lactic Acid System

Glycogen

Glucose → Glycolysis → ATP

Lactic Acid
(produced when no oxygen present)

Pyruvic Acid

Aerobic System

Oxidative Phosphorylation

Mitochondrion

MUSCLE PHYSIOLOGY / Types of Fibers

Skeletal muscle fibers vary in color depending on the myoglobin content of their cytoplasm; myoglobin is a protein that stores oxygen.

White Muscle Fibers : muscle fibers with a low concentration of myoglobin.

Red Muscle Fibers : muscle fibers with a high concentration of myoglobin; they also have more mitochondria and more blood capillaries than white muscle fibers.

There are three basic types of skeletal muscle fibers : slow oxidative, fast oxidative, and fast glycolytic. Although most skeletal muscles are a mixture of the three types, the proportions vary depending upon the action of the muscle. Postural muscles have a high proportion of slow oxidative fibers; leg muscles (used for both walking and running) have a high proportion of both slow and fast oxidative fibers; muscles used to produce large amounts of tension for short periods of time (lifting and throwing) have a high proportion of fast glycolytic fibers.

In a motor unit all fibers are of the same type.

Slow Oxidative Fibers (Type I, Slow-Twitch, or Fatigue-Resistant Fibers)

ATPase : splits ATP at a slow rate, so fibers have a slow contraction velocity.

Energy Production : ATP produced by oxidative phosphorylation (aerobic system).

Fatigue : highly resistant to fatigue; supply of oxygen and glucose keep up with the demand.

Mitochondria : contain many; sites of ATP production by oxidative phosphorylation.

Blood Capillaries : contain many; supply oxygen and glucose for oxidative phosphorylation.

Myoglobin : contain many of these oxygen-storing proteins; gives red color to fibers.

Glycogen : very little stored in these fibers.

Diameter : relatively small diameter; as in the leg muscles of long distance runners.

Location : postural muscles; for example, neck muscles that hold head up.

Fast Oxidative Fibers (Type II A, Fast-Twitch A, or Fatigue-Resistant Fibers)

ATPase : splits ATP at a fast rate, so fibers have a fast contraction velocity.

Energy Production : ATP produced by oxidative phosphorylation (aerobic system).

Fatigue : resistant to fatigue, but not as much as slow oxidative fibers.

Mitochondria : contain many; sites of ATP production by oxidative phosphorylation.

Blood Capillaries : contain many; supply oxygen and glucose for oxidative phosphorylation.

Myoglobin : contain many of these oxygen-storing proteins; gives red color to fibers.

Glycogen : very little stored in these fibers.

Diameter : intermediate in diameter; as in the leg muscles of sprinters.

Location : leg muscles of sprinters.

Fast Glycolytic Fibers (Type II B, Fast-Twitch B, or Fatigable Fibers)

ATPase : splits ATP at a fast rate; contraction is strong and rapid.

Energy Production : ATP produced by glycolysis (anaerobic process).

Fatigue : fatigue easily; when glycogen storage is depleted.

Mitochondria : contain relatively few; not used by anaerobic processes (glycolysis).

Blood Capillaries : contain relatively few; oxygen supply not needed for glycolysis.

Myoglobin : contain very few of these oxygen-storing proteins; fibers have a white color.

Glycogen : contain large amounts; ready source of fuel for anaerobic ATP production.

Diameter : large in diameter; as in the arm and leg muscles of weight lifters.

Location : arm muscles contain many of these fibers.

SKELETAL MUSCLE FIBER TYPES
Cross Section

Slow Oxidative Fibers
ATP Production : oxidative phosphorylation
Contraction Speed : slow
ATPase Activity : slow
Myoglobin Concentration : high
Mitochondria : many
Capillaries : many
Endurance : high
Diameter : small
Fiber color : red

Fast Oxidative Fibers
ATP Production : oxidative phosphorylation
Contraction Speed : fast
ATPase Activity : fast
Myoglobin Concentration : high
Mitochondria : many
Capillaries : many
Endurance : intermediate
Diameter : intermediate
Fiber color : red

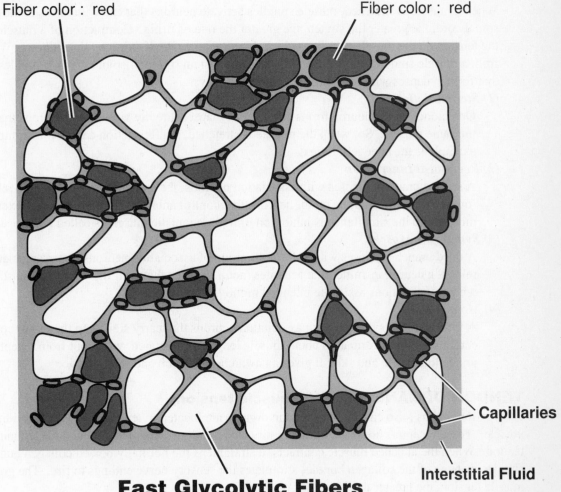

Capillaries

Interstitial Fluid

Fast Glycolytic Fibers
ATP Production : anaerobic glycolysis
Contraction Speed : fast
ATPase Activity : fast
Myoglobin Concentration : low
Mitochondria : few
Capillaries : few
Endurance : low
Diameter : large
Fiber color : white

MUSCLE PHYSIOLOGY / Proprioception

Proprioception : the awareness of the precise position of body parts.
Kinesthesia : the awareness of the directions of movement.

MUSCLE SPINDLES (monitor muscle length)

Changes in muscle length are monitored by stretch receptors called muscle spindles, which are embedded within skeletal muscle.

Anatomy

Capsule : a spindle-shaped connective tissue capsule.
Extrafusal Muscle Fibers : normal muscle fibers surrounding the capsule.
Intrafusal Muscle Fibers : 3 to 10 specialized muscle fibers enclosed in the capsule.

Nerve Pathways

When a muscle is stretched, muscle spindles activate neurons that carry impulses to the brain or spinal cord; the greater the stretch, the greater the rate of firing. Contraction of a muscle decreases the tension on the muscle spindles, which decreases the rate of firing by the neurons. When a neuron from a muscle spindle enters the central nervous system (brain or spinal cord), it divides into branches that form synapses with four different nerve pathways.

(1) Stretch Reflex

One branch of the neuron forms excitatory synapses directly with motor neurons that innervate the same muscle. So, when the muscle is stretched, a reflex action causes it to contract. An example is the knee jerk reflex.

(2) Reciprocal Innervation

A second branch synapses with association neurons; the association neurons then release neurotransmitter that inhibits motor neurons that control antagonistic muscles. The excitation of one muscle and the simultaneous inhibition of its antagonistic muscle is called reciprocal innervation.

(3) Synergistic Muscles

A third branch synapses with a different group of association neurons; these association neurons release a neurotransmitter that activates motor neurons that control synergistic muscles (muscles whose contractions assist the intended motion.)

(4) Motor Cortex

A fourth branch synapses with association neurons that carry signals to the motor cortex of the cerebrum. The information about muscle length is integrated with input from receptors in the joints, ligaments, and skin; it gives an awareness of limb and joint position.

TENDON ORGANS (monitor muscle tension)

Tendon organs (also called Golgi tendon organs) are located in tendons near the junction with its muscle. These receptors are made up of sensory nerve endings wrapped around collagen bundles of the tendon. When the attached muscle contracts, it straightens the normally bowed collagen bundles; the change in shape of the collagen bundles stimulates the sensory nerve endings to fire. The greater the muscle tension, the greater the frequency of firing.

JOINT KINESTHETIC RECEPTORS

There are 3 types of receptors located within and around articular capsules of synovial joints:

(1) Encapsulated Receptors (similar to Type II cutaneous mechanoreceptors) : respond to pressure.
(2) Small Lamellated Corpuscles (Pacinian corpuscles) : respond to acceleration and deceleration of joint movement.
(3) Tendon Organs : adjust reflex inhibition of adjacent muscles when excessive strain is placed on a joint.

TENDON ORGAN and MUSCLE SPINDLE

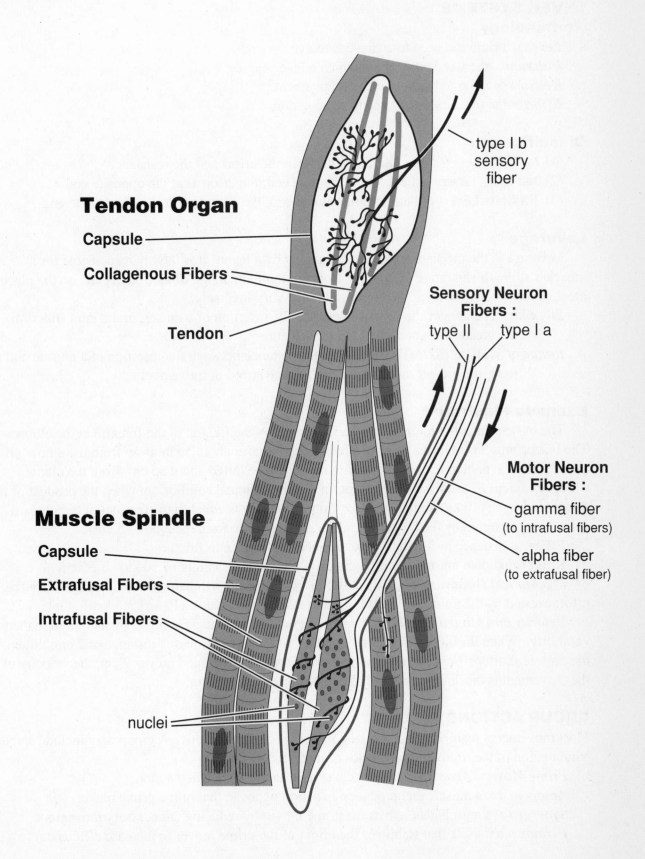

Tendon Organ

Capsule

Collagenous Fibers

Tendon

type I b
sensory
fiber

**Sensory Neuron
Fibers :**

type II   type I a

**Motor Neuron
Fibers :**

gamma fiber
(to intrafusal fibers)

alpha fiber
(to extrafusal fiber)

Muscle Spindle

Capsule

Extrafusal Fibers

Intrafusal Fibers

nuclei

MUSCLE PHYSIOLOGY / Movement Production

LEVER SYSTEMS
Terminology
Lever : a rigid rod used to achieve leverage.

Fulcrum : the fixed point about which a lever moves.

Resistance : the force that opposes movement.

Effort : the force exterted to achieve an action.

Classification
(1) 1st Class Levers : the fulcrum is between the effort and the resistance.

(2) 2nd Class Levers : the fulcrum is at one end; the effort is at the opposite end.

(3) 3rd Class Levers : the fulcrum is at one end; the resistance is at the opposite end.

Leverage
Leverage is the mechanical advantage gained by a lever. It is largely responsible for a muscle's strength and range of motion. Both strength and range of motion depend on the placement of muscle attachments. Strength and range vary inversely.

Strength The greater the distance between the insertion of a muscle and a joint (fulcrum), the greater the strength of the movement.

Range of Motion (ROM) The shorter the distance between the insertion of a muscle and a joint, the greater the range of motion and speed of movement.

Example Illustrated
The basic principle of a lever can be illustrated by the flexion of the forearm at the elbow. The biceps muscle exerts an upward force on the forearm about 5 cm away from the elbow. If a 10 kg weight is held in the hand, a downward force is exerted about 35 cm from the elbow.

Mechanical Equilibrium The forearm is in mechanical equilibrium when the product of the downward force (10 kg) and its distance from the elbow is equal to the product of the upward force (X kg) exerted by the biceps muscle and its distance from the elbow (5 cm).

Effort x distance to fulcrum = Resistance x distance to fulcrum

Effort (unknown muscle force) x 5 cm = Resistance (weight of 10 kg) x 35 cm

Mechanical Disadvantage The system is working at a mechanical disadvantage since the effort exerted by the muscle (70 kg) is greater than the resistance (10 kg).

Velocity and Maneuverability The mechanical disadvantage is offset by increased maneuverability. When the biceps shortens 1 cm, the hand moves through a distance of 7 cm. Since the muscle shortens 1 cm in the same amount of time that the hand moves 7 cm, the velocity of the movement is much greater than the rate of muscle shortening.

GROUP ACTIONS
Most movements require several skeletal muscles acting in groups. A group may include some combination of the following categories of muscles:

Prime Mover (Agonist) : the muscle that produces the desired action.

Antagonist : a muscle that produces an action opposite that of the prime mover.

Synergist : a muscle that assists the prime mover by reducing unnecessary movement.

Fixator : a muscle that stabilizes the origin of the prime mover to increase efficiency.

LEVER SYSTEMS

Antagonistic Muscles

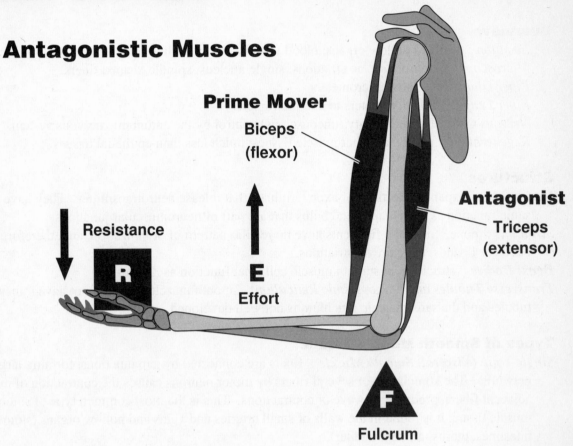

Prime Mover
Biceps
(flexor)

Antagonist
Triceps
(extensor)

Resistance

R

E
Effort

F
Fulcrum

Lever Mechanisms

Amplification

Effort

7 cm

1 cm

Distance & Velocity Amplified :
muscle shortens 1 cm
hand moves 7 cm

Inefficiency

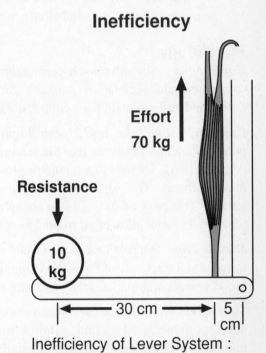

Effort
70 kg

Resistance

10
kg

30 cm

5
cm

Inefficiency of Lever System :
70 kg of muscle force required
to hold a 10 kg weight

MUSCLE PHYSIOLOGY / Smooth Muscle Tissue

Overview

Location : walls of hollow organs; blood vessels; iris and ciliary muscles; arrector pili (hair).

Microscopic Appearance : no striations; single nucleus; spindle-shaped fibers.

Fiber Diameter : 3 to 8 micrometers.

Fiber Length : 30 micrometers to 200 micrometers.

Nervous Control : involuntary (unconscious) control by the autonomic nervous system.

Regeneration : more than other muscle tissues; much less than epithelial tissues.

Structures

Varicosities : expanded portions of axon terminals that release neurotransmitter. They have the same function as the synaptic end bulbs that are part of neuromuscular junctions.

Striations : none. Since the filaments have no regular pattern of organization, and, therefore, no A bands or I bands, there are no striations.

Dense Bodies : structures in smooth muscle cells that function as Z-discs.

Transverse Tubules and Sarcoplasmic Reticulum : smooth muscle fibers do not have transverse tubules and the sarcoplasmic reticulum is not well developed.

Types of Smooth Muscle Tissue

Single-Unit (Visceral) Smooth Muscle : fibers are connected by gap junctions, forming large networks. The stimulation of several fibers by motor neurons causes the contraction of many adjacent fibers, producing a wave of contractions. This is the most common type of smooth muscle tissue; it is found in the walls of small arteries and veins and hollow organs (stomach, intestines, uterus, urinary bladder).

Multiunit Smooth Muscle : fibers act independently. Each fiber is directly stimulated by the axon terminal of a motor neuron branch. It is found in the walls of large arteries, bronchioles (airways), arrector pili muscles attached to hair follicles, radial and circular muscles of the iris that control pupil diameter, and ciliary muscles that control the shape of the lens (in the eye).

Physiology

Contraction   Smooth muscle contraction starts more slowly and lasts longer than contraction of striated muscle (skeletal and cardiac). Smooth muscle fibers can shorten and stretch to a greater extent and still maintain their contractile function.

Calcium, Calmodulin, and Myosin Light Chain Kinase   In response to stimulation by an action potential, calcium diffuses into the sarcoplasm from the sarcoplasmic reticulum and the ECF (extracellular fluid). It binds to a regulator protein called calmodulin (as it binds to troponin in striated muscle fibers). The calmodulin-calcium complex activates an enzyme called myosin light chain kinase. This enzyme uses ATP to phosphorylate (add phosphate to) the myosin head. This activates the myosin head, allowing it to bind to actin; contraction then occurs.

Muscle Tone   Muscle tone is a constant state of partial contraction. It is important in the gastrointestinal tract, where the walls maintain a steady pressure on the lumen contents; the walls of the blood vessels maintian a steady pressure on the blood.

Stress-Relaxation Response   When smooth muscle fibers are stretched, they initially contract, developing increased tension; within a minute or so the tension decreases. This allows smooth muscle to undergo great changes in length while still retaining the ability to contract effectively. Thus, smooth muscle in the walls of blood vessels and hollow organs (stomach, urinary bladder) can stretch, but the pressure on the contents within increases very little.

SMOOTH MUSCLE

Single-Unit

Fibers connected by gap junctions form networks.
A few fibers are directly stimulated by neurons;
action potentials spread via gap junctions to many adjacent fibers.

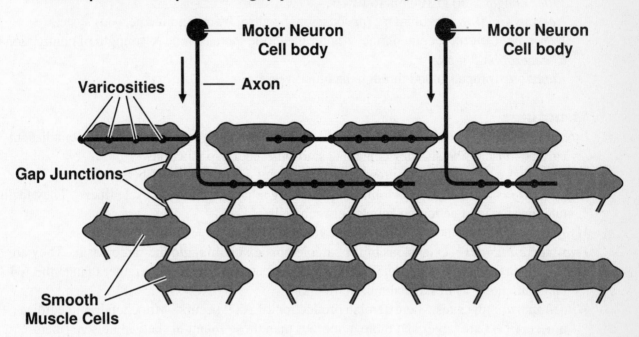

Multi-Unit

Multiunit fibers act independently.
Each fiber is directly stimulated by varicosities located on branching axons.

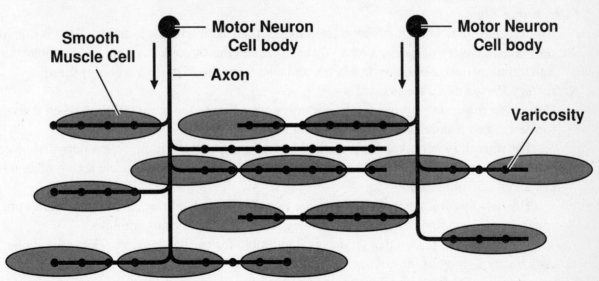

MUSCLE PHYSIOLOGY / Cardiac Muscle Tissue

Overview

Location : heart.

Microscopic Appearance : striated; single nucleus; branched fibers with intercalated discs.

Fiber Diameter : 14 micrometers.

Fiber Length : 50 to 100 micrometers.

Nervous Control : involuntary (unconscious) control by autonomic nervous system.

Hormonal Control : epinephrine & norepinephrine increase rate & strength of contractions.

Regeneration : none.

Function : propels blood through the blood vessels.

Structures

Intercalated Discs : irregular transverse thickenings of the sarcolemma that connect adjacent cardiac muscle fibers. They contain desmosomes and gap junctions.

Desmosomes : rivet-like structures that hold adjacent cardiac muscle fibers together.

Gap Junctions : channels of cytoplasm connecting adjacent cardiac muscle fibers. They facilitate the spread of action potentials between cells.

Myofibrils : not present in cardiac muscle fibers.

Transverse Tubules : extensions of the sarcolemma extending into the sarcoplasm. They are smaller than in skeletal muscle fibers, and are located at the Z-discs (rather than at the A-I band junctions, as in skeletal muscle fibers).

Mitochondria : the sites where aerobic production of ATP occurs. Mitochondria in cardiac muscle fibers are larger and more numerous than those found in skeletal muscle fibers.

Physiology

Autorhythmicity

Cardiac muscle fibers can contract without extrinsic stimulation from nerves or hormones. The cells in the pacemaker spontaneously discharge about 100 times per minute. Nervous and hormonal input can alter the rate of this automatic discharge. The slowing effect of the parasympathetic nerves is predominant; the normal heart rate is about 70 beats per minute.

Contraction Time

Cardiac muscle fibers remain contracted longer than skeletal muscle fibers. This is due to the prolonged delivery of calcium ions. Calcium enters the sarcoplasm from the sarcoplasmic reticulum (as in skeletal muscle fibers), and also from the ECF (extracellular fluid).

Refractory Period and Tetanus

Refractory means unresponsive. In physiology it refers to the period of time when a muscle (or nerve) cell is unresponsive to stimulation.

The refractory period of a muscle cell lasts as long as the muscle action potential. Skeletal muscle action potentials last only 1 millisecond, while cardiac muscle action potentials last about 250 milliseconds.

This is why summation and tetanus are possible in skeletal muscle tissue; muscles are capable of responding to stimuli before they relax from a previous stimulus.

This is also the reason that tetanus is impossible for cardiac muscle. The action potential lasts nearly as long as the contraction and relaxation periods, so a cardiac muscle cell will have nearly completed its relaxation period before it can respond to a subsequent stimulus. If tetanus could occur in cardiac muscle, it would stop the heartbeat and the flow of blood.

REFRACTORY PERIOD

The refractory period lasts as long as the action potential.
During this time the muscle membrane cannot respond to stimuli.

Skeletal Muscle

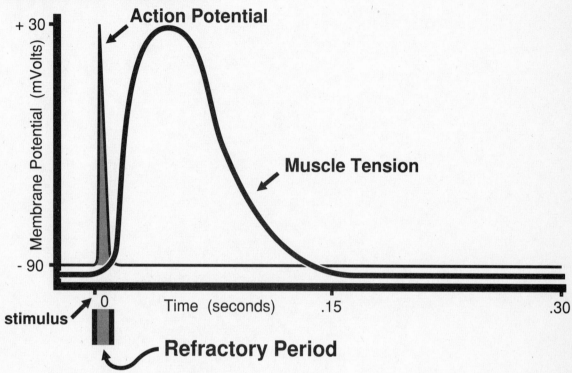

Cardiac Muscle

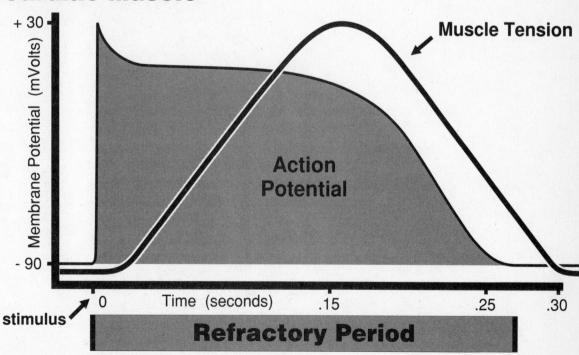

3 Skeletal Muscles
(major superficial muscles)

SKELETAL MUSCLES / Naming Skeletal Muscles

There are about 700 skeletal muscles. They are named on the basis of distinctive criteria : size, shape, location, action, origin and insertion, number of origins (or heads), and direction of muscle fibers.

Origin : the end of a muscle that attaches to a bone that remains stationary during muscle contraction.

Insertion : the end of a muscle that attaches to a bone that moves during muscle contraction.

Size
Maximus = largest.   *Example :* gluteus maximus.
Minimus = smallest.   *Example :* gluteus minimus.
Longus = longest.   *Example :* adductor longus.
Brevis = short.   *Example :* peroneus brevis.

Shape
Deltoid = triangular.   *Example :* deltoid.
Trapezius = trapezoid.   *Example :* trapezius.
Serratus = saw-toothed.   *Example :* serratus anterior.
Rhomboideus = diamond-shaped.   *Example :* rhomboideus major.

Location
Example : temporalis (near the temporal bone).
Example : tibialis anterior (near the front of the tibia).
Example : orbicularis oculi (surrounding the eye).
Example : orbicularis oris (surrounding the mouth).

Action
Flexor : decreases the angle at a joint.   *Example :* flexor carpi radialis.
Extensor : increases the angle at a joint.   *Example :* extensor carpi ulnaris.
Abductor : moves a bone away from the midline.   *Example :* abductor longus.
Adductor : moves a bone closer to the midline.   *Example :* adductor longus.
Levator : produces an upward movement.   *Example :* levator ani.
Depressor : produces a downward movement.   *Example :* depressor labii inferioris.
Supinator : turns the palm upward or anteriorly.   *Example :* supinator.
Pronator : turns the palm downward or posteriorly.   *Example :* pronator teres.
Sphincter : decreases the size of an opening.   *Example :* external anal sphincter.
Tensor : makes a body part more rigid.   *Example :* tensor fasciae latae.
Rotator : moves a bone around its longitudinal axis.   *Example :* obturator externus.

Origin and Insertion
Example : sternocleidomastoid (*origin :* sternum and clavicle; *insertion :* mastoid process).
Example : sternohyoid (*origin :* sternum; *insertion :* hyoid bone).

Number of Origins
Biceps : two origins.   *Examples :* biceps brachii; biceps femoris.
Triceps : three origins.   *Example :* triceps brachii.
Quadriceps : four origins.   *Example :* quadriceps femoris.

Direction of Muscle Fibers
Rectus : fibers run parallel to the midline of the body.   *Example :* rectus abdominis.
Transverse : fibers run perpendicular to the midline.   *Example :* transverse abdominis.
Oblique : fibers run diagonally to the midline.   *Example :* external oblique.

MAJOR MUSCLES : Anterior View

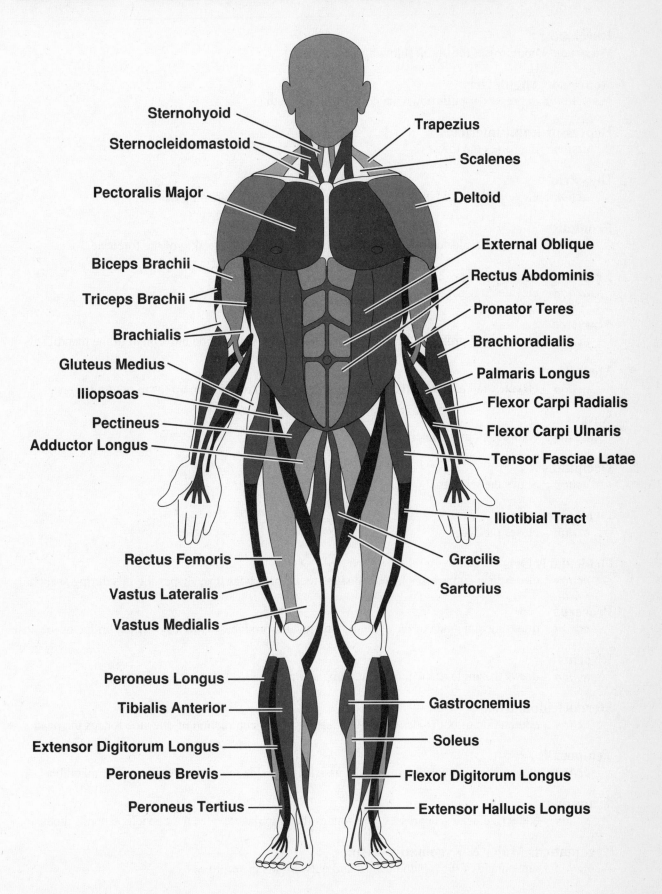

Sternohyoid

Sternocleidomastoid

Pectoralis Major

Biceps Brachii

Triceps Brachii

Brachialis

Gluteus Medius

Iliopsoas

Pectineus

Adductor Longus

Rectus Femoris

Vastus Lateralis

Vastus Medialis

Peroneus Longus

Tibialis Anterior

Extensor Digitorum Longus

Peroneus Brevis

Peroneus Tertius

Trapezius

Scalenes

Deltoid

External Oblique

Rectus Abdominis

Pronator Teres

Brachioradialis

Palmaris Longus

Flexor Carpi Radialis

Flexor Carpi Ulnaris

Tensor Fasciae Latae

Iliotibial Tract

Gracilis

Sartorius

Gastrocnemius

Soleus

Flexor Digitorum Longus

Extensor Hallucis Longus

83

Buccinator
action : compresses the cheek (blowing and sucking).

Depressor Anguli Oris
action : depresses or pulls down the corner of the mouth.

Depressor Labii Inferioris
action : depresses the lower lip.

Digastric
action : elevates the hyoid bone; depresses the mandible (as in opening the mouth).

Frontalis
action : draws the scalp forward; raises the eyebrows; wrinkles the skin of the forehead.

Levator Labii Superioris
action : elevates (raises) the upper lip.

Masseter
action : elevates and protracts the mandible; assists in the side to side movement of the mandible.

Mentalis
action : elevates and protrudes the lower lip; pulls the skin of the chin up (as in pouting).

Nasalis
action : compresses and dilates the nostrils.

Occipitalis
action : draws the scalp backward.

Orbicularis Oculi
action : closes the eye.

Orbicularis Oris
action : closes lips; compresses lips against the teeth; protrudes lips; shapes the lips during speech.

Procerus
action : draws medial angle of the eyebrow downward; produces wrinkles over the bridge of the nose.

Risorius
action : draws the angle of the mouth laterally, as in tenseness.

Sternocleidomastoid
action : contraction of both sides draws the head forward; contraction of one side rotates the head.

Temporalis
action : elevates and retracts the mandible; assists in the side to side movement of the mandible.

Trapezius
action : adducts, elevates, depresses, and rotates the scapula; elevates the clavicle; extends the head.

Zygomaticus Major & Zygomaticus Minor
action : draw angle of the mouth upward and outward (as in smiling).

HEAD and NECK (lateral view)

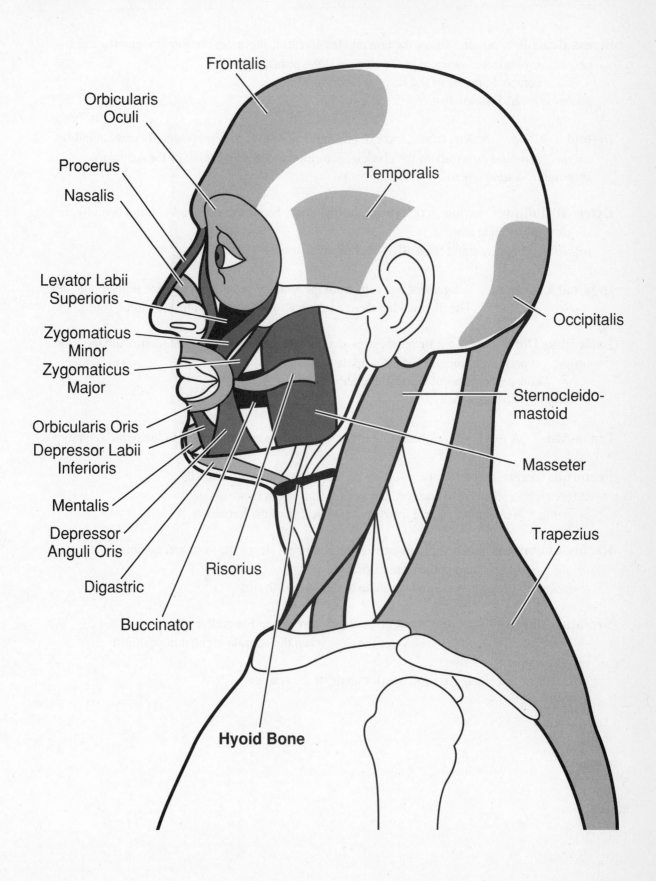

Frontalis

Orbicularis Oculi

Procerus

Nasalis

Temporalis

Levator Labii Superioris

Zygomaticus Minor

Zygomaticus Major

Occipitalis

Orbicularis Oris

Depressor Labii Inferioris

Sternocleido-mastoid

Mentalis

Depressor Anguli Oris

Masseter

Digastric

Risorius

Trapezius

Buccinator

Hyoid Bone

Biceps Brachii   *action :* flexes the arm and the forearm; supinates the forearm and the hand.
   origin : supraglenoid tubercle of the scapula (long head);
            coracoid process of the scapula (short head).
   insertion : radial tuberosity.

Deltoid   *action :* abducts, flexes, and extends the arm; rotates the arm (medially and laterally).
   origin : acromial extremity of the clavicle; acromion process and spine of the scapula.
   insertion : deltoid tuberosity of the humerus.

External Oblique   *action :* compresses the abdomen; bends the vertebral column laterally.
   origin : lower eight ribs.
   insertion : iliac crest and linea alba (midline aponeurosis).

Inguinal Ligament   A ligament that runs from the anterior superior iliac spine to the pubic tubercle.
            The inguinal ligament demarcates the thigh and the abdominal wall.

Latissimus Dorsi   *action :* flexes the arm and the forearm; supinates the forearm and the hand.
   origin : supraglenoid tubercle of the scapula (long head);
            coracoid process of the scapula (short head).
   insertion : radial tuberosity.

Linea Alba   A tough, fibrous band that extends from the xiphoid process to the pubic symphysis.

Pectoralis Major   *action :* flexes, adducts, and rotates the arm medially.
   origin : clavicle, sternum and cartilages of the 2nd to 6th ribs.
   insertion : greater tubercle and intertubercular sulcus of the humerus.

Rectus Abdominis   *action :* compresses the abdomen; flexes the vertebral column.
   origin : pubic crest and pubic symphysis.
   insertion : xiphoid process and cartilage of the 5th to 7th ribs.

Serratus Anterior   *action :* rotates the scapula upward and laterally;
                     elevates the ribs (when the scapula is fixed in position).
   origin : upper 8 or 9 ribs
   insertion : vertebral border and inferior angle of the scapula.

TORSO (anterior view)

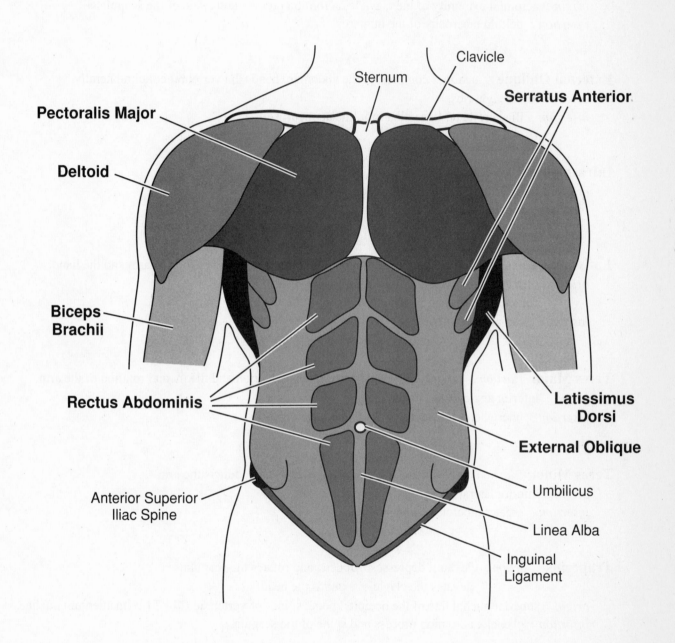

Sternum

Clavicle

Serratus Anterior

Pectoralis Major

Deltoid

Biceps Brachii

Rectus Abdominis

Latissimus Dorsi

External Oblique

Umbilicus

Anterior Superior Iliac Spine

Linea Alba

Inguinal Ligament

SKELETAL MUSCLES / Torso (posterior view)

Deltoid   *action :* abducts, flexes, and extends the arm; rotates the arm (medially and laterally).
  origin : acromial extremity of the clavicle; acromion process and spine of the scapula.
  insertion : deltoid tuberosity of the humerus.

External Oblique   *action :* compresses the abdomen; bends the vertebral column laterally.
  origin : lower eight ribs.
  insertion : iliac crest and linea alba (midline aponeurosis).

Infraspinatus   *action :* rotates the arm laterally; adducts the arm.
  origin : fossa inferior to the spine of the scapula.
  insertion : greater tubercle of the humerus.

Latissimus Dorsi   *action :* flexes the arm and the forearm; supinates the forearm and the hand.
  origin : supraglenoid tubercle of the scapula (long head);
        coracoid process of the scapula (short head).
  insertion : radial tuberosity.

Teres Major   *action :* extends the arm; assists in the adduction and the medial rotation of the arm.
  origin : inferior angle of the scapula.
  insertion : intertubercular sulcus of the humerus.

Teres Minor   *action :* rotates the arm laterally; extends and adducts the arm.
  origin : inferior lateral border of the scapula.
  insertion : greater tubercle of the humerus.

Trapezius   *action :* elevates, depresses, adducts, and rotates the scapula;
            elevates the clavicle; extends the head.
  origin : superior nuchal line of the occipital bone; spines of vertebrae C7 - T12; ligamentum nuchae.
  insertion : clavicle; acromion process and spine of the scapula.

Triceps Brachii   *action :* extends the forearm; extends the arm.
  origin :
      long head : infraglenoid tuberosity of the scapula.
      lateral head : lateral and posterior surface of the humerus superior to the radial groove.
      medial head : posterior surface of the humerus inferior to the radial groove.
  insertion : olecranon of the ulna.

TORSO (posterior view)

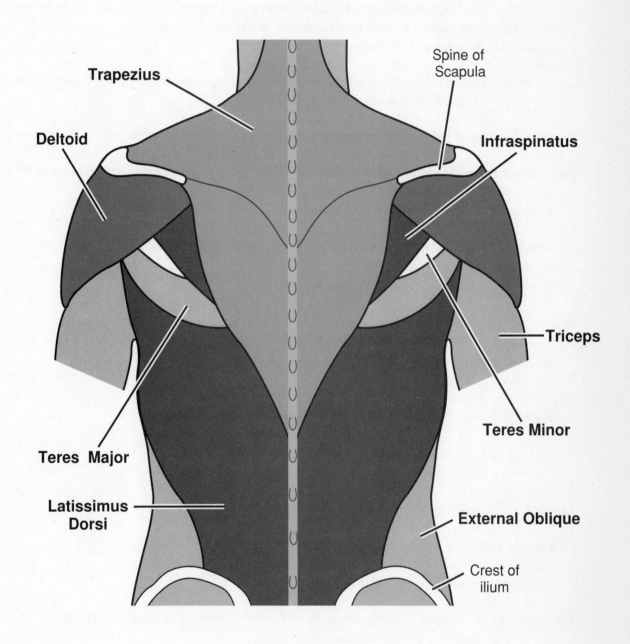

Trapezius

Deltoid

Spine of
Scapula

Infraspinatus

Triceps

Teres Minor

Teres Major

Latissimus
Dorsi

External Oblique

Crest of
ilium

Anococcygeal Raphe   The line or ridge of tissue connecting the anus with the coccyx.

Bulbocavernosus
action :
     male : helps expel urine during urination; helps propel semen; assists in erection of the penis.
     female : decreases the vaginal orifice; assists in erection of the clitoris.
origin : central tendon of the perineum.
insertion :
     male : inferior fascia of urogenital diaphragm; corpus spongiosum and deep fascia of penis.
     female : pubic arch; root and dorsum of clitoris.

Central Tendon of Perineum   Tendon connecting external anal sphincter with bulbocavernosus.

Deep Transverse Perineus   *action :* helps expel last drops of urine; helps propel semen in the male.
origin : ischial rami.
insertion : central tendon of the perineum.

External Anal Sphincter   *action :* keeps the anal canal and the orifice closed.
origin : anococcygeal raphe.
insertion : central tendon of the perineum.

Iliococcygeus   *action :* slightly raises the pelvic floor; resists intra-abdominal pressure;
                            constricts the anus.
origin : ischial spine.
insertion : coccyx.

Ischiocavernosus   *action :* may maintain the erection of the penis in the male
                          and the erection of the clitoris in the female.
origin : ischial tuberosity; ischial and pubic rami.
insertion : corpus cavernosum of the penis in the male; clitoris in the female.

Levator Ani   consists of 2 muscles : iliococcygeus and pubococcygeus.

Obturator Internus   *action :* abducts the thigh; rotates the thigh laterally.
origin : inner surface of the obturator foramen, pubis, and ischium.
insertion : greater trochanter of the femur.

Pubococcygeus   *action :* slightly raises the pelvic floor; resists intra-abdominal pressure;
                          constricts the anus.
origin : pubis.
insertion : coccyx; urethra; anal canal; central tendon of the perineum; anococcygeal raphe.

Superficial Transverse Perineus   *action :* helps to stabilize the central tendon of the perineum.
origin : ischial tuberosity.
insertion : central tendon of the perineum.

PERINEUM

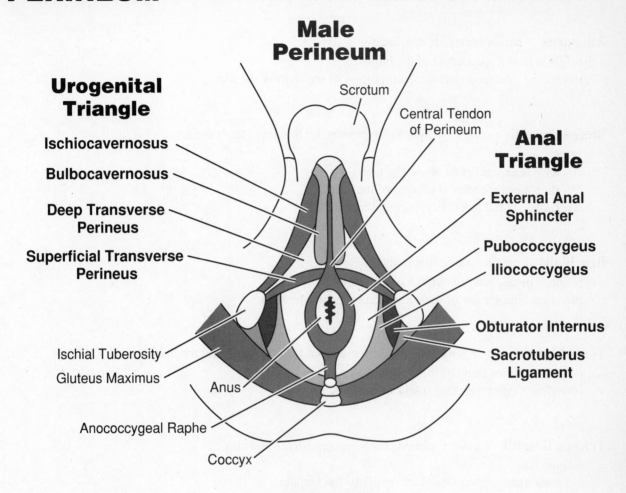

Male Perineum

Urogenital Triangle

- Ischiocavernosus
- Bulbocavernosus
- Deep Transverse Perineus
- Superficial Transverse Perineus
- Ischial Tuberosity
- Gluteus Maximus
- Anus
- Anococcygeal Raphe
- Coccyx

Scrotum

Central Tendon of Perineum

Anal Triangle

- External Anal Sphincter
- Pubococcygeus
- Iliococcygeus
- Obturator Internus
- Sacrotuberus Ligament

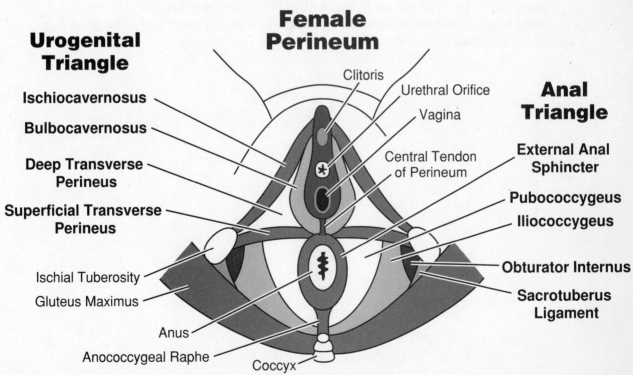

Female Perineum

Urogenital Triangle

- Ischiocavernosus
- Bulbocavernosus
- Deep Transverse Perineus
- Superficial Transverse Perineus
- Ischial Tuberosity
- Gluteus Maximus
- Anus
- Anococcygeal Raphe
- Coccyx

Clitoris

Urethral Orifice

Vagina

Central Tendon of Perineum

Anal Triangle

- External Anal Sphincter
- Pubococcygeus
- Iliococcygeus
- Obturator Internus
- Sacrotuberus Ligament

SKELETAL MUSCLES / Arm (anterior and posterior views)

Anconeus   *action :* extends the forearm.
  origin : lateral epicondyle of the humerus.
  insertion : olecranon and superior portion of the shaft of the ulna.

Biceps Brachii   *action :* flexes and supinates the forearm; flexes the arm.
  origin :
        long head : tubercle above the glenoid cavity.
        short head : coracoid process of the scapula.
  insertion : radial tuberosity; bicipital aponeurosis.

Brachialis   *action :* flexes the forearm.
  origin : distal, anterior surface of the humerus.
  insertion : tuberosity of the coronoid process of the ulna.

Teres Major   *action :* extends the arm; assists in the adduction and medial rotation of the arm.
  origin : inferior angle of the scapula.
  insertion : intertubercular sulcus of the humerus.

Triceps Brachii   *action :* extends the forearm; extends the arm.
  origin :
        long head : infraglenoid tuberosity of the scapula.
        medial head : posterior surface of the humerus inferior to the radial groove.
        lateral head : lateral and posterior surface of the humerus superior to the radial groove.
  insertion : olecranon of the ulna.

ARM (anterior and posterior views)

Anterior View　　　　　Posterior View

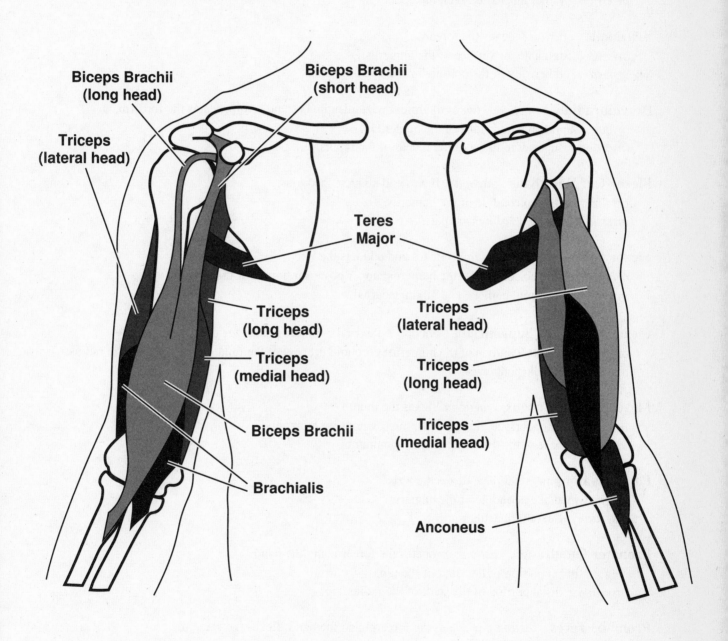

Biceps Brachii
(long head)

Biceps Brachii
(short head)

Triceps
(lateral head)

Teres
Major

Triceps
(long head)

Triceps
(lateral head)

Triceps
(medial head)

Triceps
(long head)

Biceps Brachii

Triceps
(medial head)

Brachialis

Anconeus

Biceps Brachii *action :* flexes and supinates the forearm; flexes the arm.
   origin :
      long head : tubercle above the glenoid cavity.
      short head : coracoid process of the scapula.
   insertion : radial tuberosity; bicipital aponeurosis.

Brachialis *action :* flexes the forearm.
   origin : distal, anterior surface of the humerus.
   insertion : tuberosity of the coronoid process of the ulna.

Brachioradialis *action :* flexes the forearm; semisupinates and semipronates the forearm.
   origin : supracondyloid ridge of the humerus.
   insertion : superior to the styloid process of the radius.

Flexor Carpi Radialis *action :* flexes and abducts the wrist.
   origin : medial epicondyle of the humerus.
   insertion : 2nd and 3rd metacarpals.

Flexor Carpi Ulnaris *action :* flexes and adducts the wrist.
   origin : medial epicondyle of the humerus; upper posterior border of the ulna.
   insertion : pisiform, hamate, and 5th metacarpal.

Flexor Digitorum Superficialis *action :* flexes the middle phalanges of each finger.
   origin : medial epicondyle of the humerus; coronoid process of the ulna; oblique line of the radius.
   insertion : middle phalanges.

Flexor Pollicis Longus *action :* flexes the thumb.
   origin : anterior surface of the radius; interosseous membrane.
   insertion : base of the distal phalanx of the thumb.

Palmaris Longus *action :* flexes the wrist.
   origin : medial epicondyle of the humerus.
   insertion : flexor retinaculum.

Pronator Quadratus *action :* pronates the forearm and the hand.
   origin : distal portion of the shaft of the ulna.
   insertion : distal portion of the shaft of the radius.

Pronator Teres *action :* pronates the forearm and the hand; flexes the forearm.
   origin : medial epicondyle of the humerus; coronoid process of the ulna.
   insertion : midlateral surface of the radius.

Retinacula (*retinere* = retain) At the wrist, the deep fascia is thickened into fibrous bands called retinacula (singular : retinaculum). The ***flexor retinaculum*** (*transverse carpal ligament*) is located over the palmar surface of the carpal bones; the ***extensor retinaculum*** (*dorsal carpal ligament*) is located over the dorsal surface of the carpal bones.

FOREARM (anterior view)

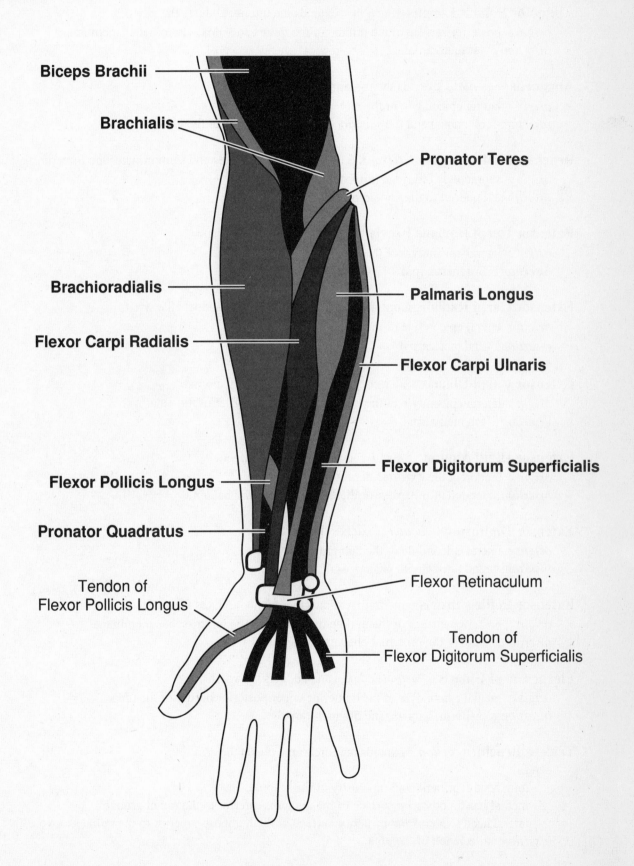

Biceps Brachii

Brachialis

Pronator Teres

Brachioradialis

Palmaris Longus

Flexor Carpi Radialis

Flexor Carpi Ulnaris

Flexor Digitorum Superficialis

Flexor Pollicis Longus

Pronator Quadratus

Tendon of
Flexor Pollicis Longus

Flexor Retinaculum

Tendon of
Flexor Digitorum Superficialis

Abductor Pollicis Longus   *action :* extends the thumb; abducts the wrist.
    origin : posterior surface of the middle of the radius and ulna; interosseous membrane.
    insertion : 1st metacarpal.

Anconeus   *action :* extends the forearm.
    origin : lateral epicondyle of the humerus.
    insertion : olecranon and the superior portion of the shaft of the ulna.

Brachioradialis   *action :* flexes the forearm; semisupinates and semipronates the forearm.
    origin : supracondyloid ridge of the humerus.
    insertion : superior to the styloid process of the radius.

Extensor Carpi Radialis Brevis   *action :* extends the wrist.
    origin : lateral epicondyle of the humerus.
    insertion : 3rd metacarpal.

Extensor Carpi Radialis Longus   *action :* extends and abducts the wrist.
    origin : lateral epicondyle of the humerus.
    insertion : 2nd metacarpal.

Extensor Carpi Ulnaris   *action :* extends and adducts the wrist.
    origin : lateral epicondyle of the humerus; posterior border of the ulna.
    insertion : 5th metacarpal.

Extensor Digiti Minimi   *action :* extends the little finger.
    origin : tendon of the extensor digitorum.
    insertion : tendon of the extensor digitorum on the 5th phalanx.

Extensor Digitorum   *action :* extends the phalanges.
    origin : lateral epicondyle of the humerus.
    insertion : 2nd through 5th phalanges.

Extensor Pollicis Brevis   *action :* extends the thumb; abducts the wrist.
    origin : posterior surface of the middle of the radius; the interosseous membrane.
    insertion : base of the proximal phalanx of the thumb.

Flexor Carpi Ulnaris   *action :* flexes and adducts the wrist.
    origin : medial epicondyle of the humerus; upper posterior border of the ulna.
    insertion : pisiform, hamate, and 5th metacarpal.

Triceps Brachii   *action :* extends the forearm; extends the arm.
    origin :
       long head : infraglenoid tuberosity of the scapula.
       medial head : posterior surface of the humerus inferior to the radial groove.
       lateral head : lateral and posterior surface of the humerus superior to the radial groove.
    insertion : olecranon of the ulna.

FOREARM (posterior view)

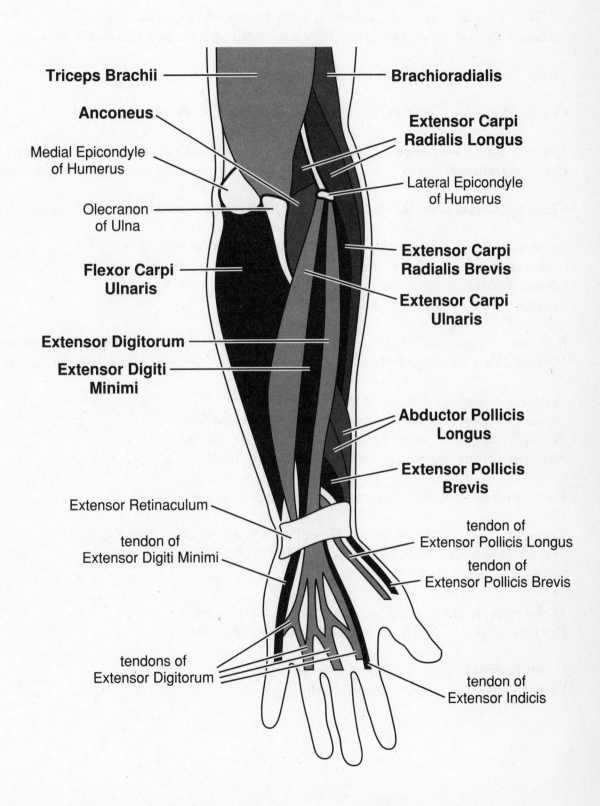

Triceps Brachii

Anconeus

Medial Epicondyle
of Humerus

Olecranon
of Ulna

Flexor Carpi
Ulnaris

Extensor Digitorum

Extensor Digiti
Minimi

Extensor Retinaculum

tendon of
Extensor Digiti Minimi

tendons of
Extensor Digitorum

Brachioradialis

Extensor Carpi
Radialis Longus

Lateral Epicondyle
of Humerus

Extensor Carpi
Radialis Brevis

Extensor Carpi
Ulnaris

Abductor Pollicis
Longus

Extensor Pollicis
Brevis

tendon of
Extensor Pollicis Longus

tendon of
Extensor Pollicis Brevis

tendon of
Extensor Indicis

SKELETAL MUSCLES / Lower Extremity

Major Superficial Muscles (alphabetical list)

Adductor Longus   *action :* adducts, laterally rotates, and flexes the thigh.
Adductor Magnus   *action :* adducts, flexes, laterally rotates, and extends the thigh.

Biceps Femoris   *action :* flexes the leg; extends the thigh.

Calcaneal Tendon   Tendon of the soleus, gastrocnemius, and plantaris muscles.

Extensor Digitorum Longus   *action :* dorsiflexes and everts the foot; extends the toes.
Extensor Hallucis Longus   *action :* dorsiflexes and everts the foot; extends the great toe.

Flexor Digitorum Longus   *action :* plantar flexes and inverts the foot; flexes the toes.

Gastrocnemius   *action :* plantar flexes the foot; flexes the leg.
Gluteus Maximus   *action :* extends and rotates the thigh laterally.
Gluteus Medius   *action :* abducts and rotates the thigh medially.
Gracilis   *action :* adducts the thigh; flexes the leg.

Iliopsoas   2 muscles : iliacus and psoas major   *action :* flexes and rotates the thigh laterally.
Iliotibial Tract   Formed by fascia lata and tendons of tensor fasciae latae and gluteus maximus.

Pectineus   *action :* flexes and adducts the thigh.
Peroneus Brevis   *action :* plantar flexes and everts the foot.
Peroneus Longus   *action :* plantar flexes and everts the foot.
Peroneus Tertius   *action :* dorsiflexes and everts the foot.

Rectus Femoris   *action :* extends the leg; flexes the thigh.

Sartorius   *action :* flexes the leg; flexes the thigh and rotates it laterally.
Semimembranosus   *action :* flexes the thigh; extends the thigh.
Soleus   *action :* plantar flexes the foot.

Tensor Fasciae Latae   *action :* flexes and abducts the thigh.
Tibialis Anterior   *action :* dorsiflexes and inverts the foot.

Vastus Lateralis   *action :* extends the leg.
Vastus Medialis   *action :* extends the leg.

LOWER EXTREMITY (lateral and anterior views)

Lateral View

Anterior View

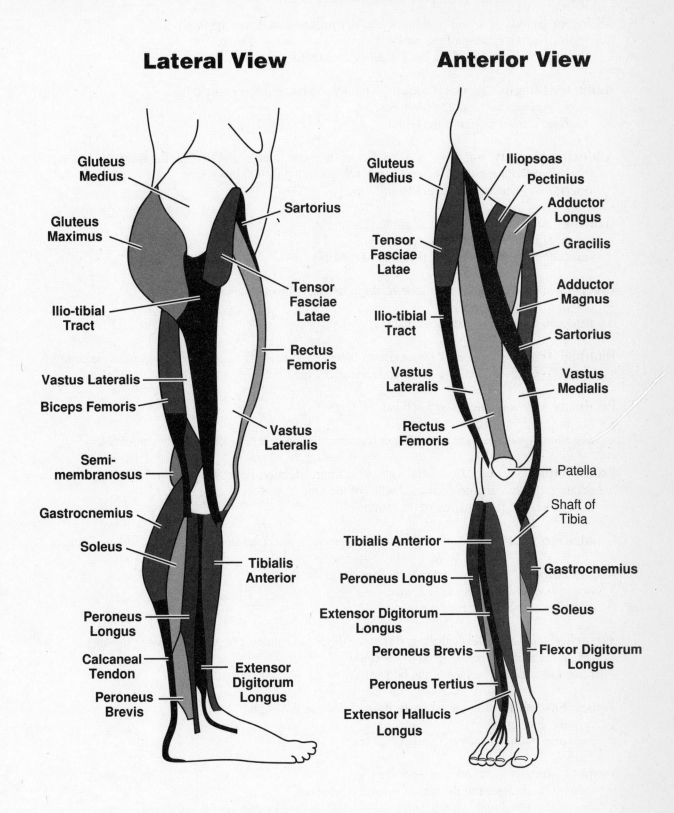

Lateral View labels:
- Gluteus Medius
- Gluteus Maximus
- Ilio-tibial Tract
- Vastus Lateralis
- Biceps Femoris
- Semi-membranosus
- Gastrocnemius
- Soleus
- Peroneus Longus
- Calcaneal Tendon
- Peroneus Brevis
- Sartorius
- Tensor Fasciae Latae
- Rectus Femoris
- Vastus Lateralis
- Tibialis Anterior
- Extensor Digitorum Longus

Anterior View labels:
- Gluteus Medius
- Tensor Fasciae Latae
- Ilio-tibial Tract
- Vastus Lateralis
- Rectus Femoris
- Tibialis Anterior
- Peroneus Longus
- Extensor Digitorum Longus
- Peroneus Brevis
- Peroneus Tertius
- Extensor Hallucis Longus
- Iliopsoas
- Pectinius
- Adductor Longus
- Gracilis
- Adductor Magnus
- Sartorius
- Vastus Medialis
- Patella
- Shaft of Tibia
- Gastrocnemius
- Soleus
- Flexor Digitorum Longus

Adductor Brevis   *action :* adducts, laterally rotates, and flexes the thigh.
   origin : inferior ramus of the pubis.
   insertion : upper half of the linea aspera of the femur.

Adductor Longus   *action :* adducts, laterally rotates, and flexes the thigh.
   origin : pubic crest; pubic symphysis.
   insertion : linea aspera of the femur.

Adductor Magnus   *action :* adducts, flexes, laterally rotates, and extends the thigh.
   origin : inferior ramus of the pubis and ischium to the ischial tuberosity.
   insertion : linea aspera of the femur.

Gracilis   *action :* adducts the thigh; flexes the leg.
   origin : pubic arch; pubic symphysis.
   insertion : medial surface of the body of the tibia.

Iliacus   *action :* flexes and rotates the thigh laterally; flexes the vertebral column.
   origin : iliac fossa.
   insertion : tendon of the psoas major.

Iliotibial Tract   Formed by fascia lata and tendons of tensor fasciae latae and gluteus maximus.
   Inserts into the lateral condyle of the tibia.

Pectineus   *action :* flexes and adducts the thigh.
   origin : superior ramus of the pubis.
   insertion : pectineal line of the femur (between the linea aspera and the lesser trochanter).

Psoas Major   *action :* flexes and rotates the thigh laterally; flexes the vertebral column.
   origin : transverse processes and bodies of the lumbar vertebrae.
   insertion : lesser trochanter of the femur.

Quadriceps Femoris   4 muscles : rectus femoris; vastus lateralis, intermedius, and medialis.

Rectus Femoris   *action :* extends the leg; flexes the thigh.
   origin : anterior inferior iliac spine.
   insertion : upper border of the patella.

Sartorius   *action :* flexes the leg; flexes the thigh and rotates it laterally (crossing the leg).
   origin : anterior superior spine of the ilium.
   insertion : medial surface of the body of the tibia.

Tensor Fasciae Latae   *action :* flexes and abducts the thigh.
   origin : iliac crest.
   insertion : tibia by way of the iliotibial tract.

Vastus Lateralis   *action :* extends the leg.
   origin : linea aspera of the femur; greater trochanter.
   insertion : tibial tuberosity through the patellar ligament (tendon of quadriceps).

Vastus Medialis   *action :* extends the leg.
   origin : linea aspera of the femur.
   insertion : tibial tuberosity through the patellar ligament (tendon of quadriceps).

THIG (anterior view)

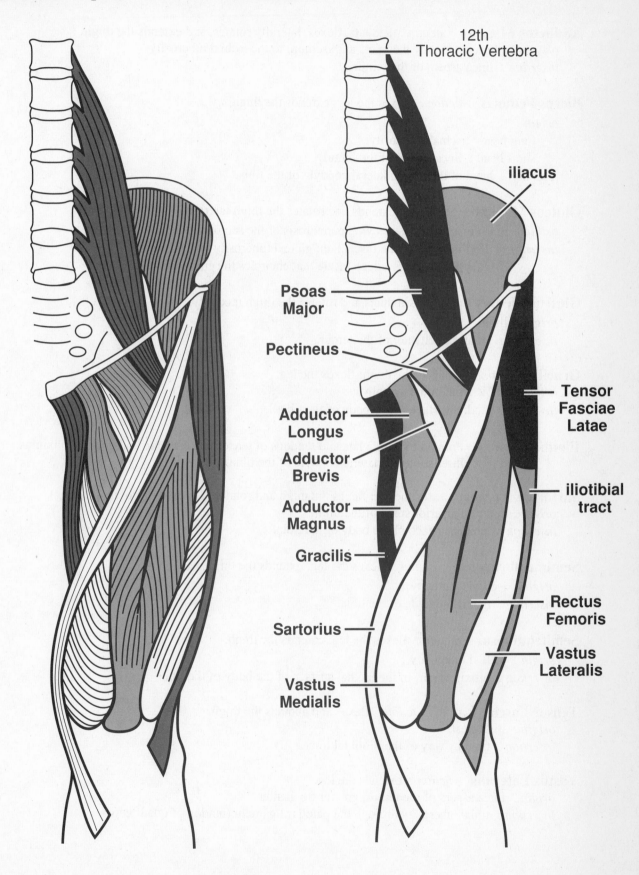

12th
Thoracic Vertebra

iliacus

Psoas
Major

Pectineus

Adductor
Longus

Adductor
Brevis

Adductor
Magnus

Gracilis

Sartorius

Vastus
Medialis

Tensor
Fasciae
Latae

iliotibial
tract

Rectus
Femoris

Vastus
Lateralis

101

SKELETAL MUSCLES / Thigh (posterior view)

Adductor Magnus *action :* adducts, flexes, laterally rotates, and extends the thigh.
origin : inferior ramus of the pubis and ischium to the ischial tuberosity.
insertion : linea aspera of the femur.

Biceps Femoris *action :* flexes the leg; extends the thigh.
origin :
>   long head : ischial tuberosity.
>   short head : linea aspera of the femur.

insertion : head of the fibula; lateral condyle of the tibia.

Gluteus Maximus *action :* extends and rotates the thigh laterally.
origin : iliac crest; sacrum; coccyx; aponeurosis of the sacrospinalis.
insertion : iliotibial tract of the fascia lata; gluteal tuberosity of the femur.
>   (the fascia lata is a deep fascia that encircles the entire thigh)

Gluteus Medius *action :* abducts and rotates the thigh medially.
origin : ilium.
insertion : greater trochanter of the femur.

Gracilis *action :* adducts the thigh; flexes the leg.
origin : pubic arch; pubic symphysis.
insertion : medial surface of the body of the tibia.

Iliotibial Tract Formed by fascia lata and tendons of tensor fasciae latae and gluteus maximus.
Inserts into the lateral condyle of the tibia.

Sartorius *action :* flexes the leg; flexes the thigh and rotates it laterally (crossing the leg).
origin : anterior superior spine of the ilium.
insertion : medial surface of the body of the tibia.

Semimembranosus *action :* flexes the leg; extends the thigh.
origin : ischial tuberosity.
insertion : medial condyle of the tibia.

Semitendinosus *action :* flexes the leg; extends the thigh.
origin : ischial tuberosity.
insertion : proximal part of the medial surface of the body of the tibia.

Tensor Fasciae Latae *action :* flexes and abducts the thigh.
origin : iliac crest.
insertion : tibia by way of the iliotibial tract.

Vastus Lateralis *action :* extends the leg.
origin : linea aspera of the femur; greater trochanter.
insertion : tibial tuberosity through the patellar ligament (tendon of quadriceps).

THIGH (posterior view)

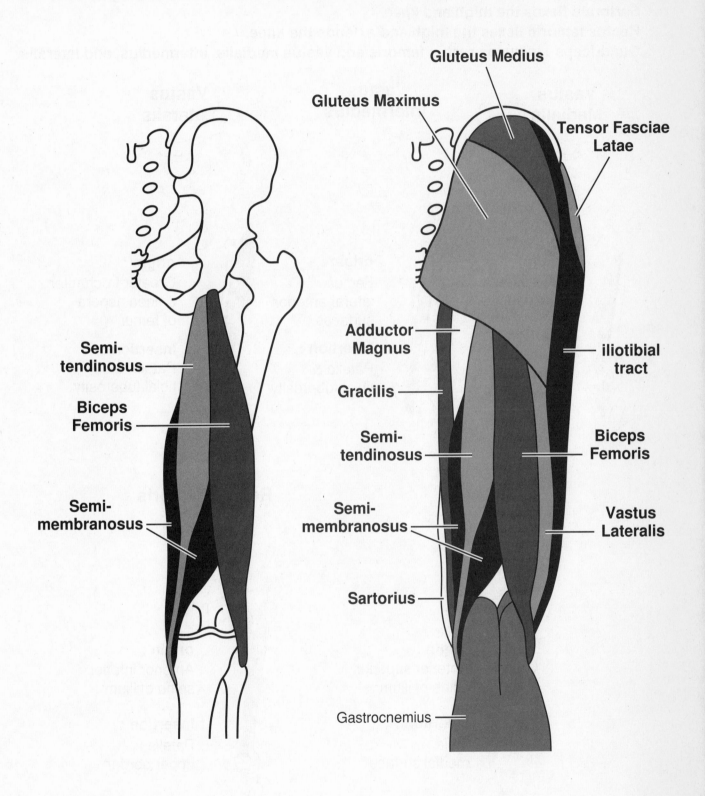

Gluteus Medius

Gluteus Maximus

Tensor Fasciae Latae

Semi-tendinosus

Biceps Femoris

Semi-membranosus

Adductor Magnus

Gracilis

Semi-tendinosus

Semi-membranosus

Sartorius

Gastrocnemius

iliotibial tract

Biceps Femoris

Vastus Lateralis

THIGH : anterior muscles

Vastus medialis, vastus intermedius, and vastus lateralis extend the leg.
Sartorius flexes the thigh and knee.
Rectus femoris flexes the thigh and extends the knee.
Quadriceps : includes rectus femoris and vastus medialis, intermedius, and lateralis.

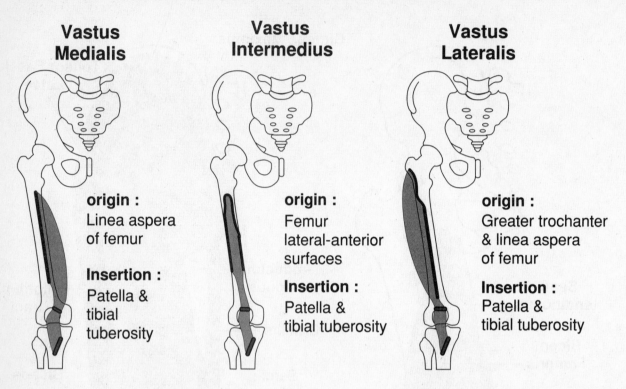

Vastus Medialis

origin :
Linea aspera
of femur

Insertion :
Patella &
tibial
tuberosity

Vastus Intermedius

origin :
Femur
lateral-anterior
surfaces

Insertion :
Patella &
tibial tuberosity

Vastus Lateralis

origin :
Greater trochanter
& linea aspera
of femur

Insertion :
Patella &
tibial tuberosity

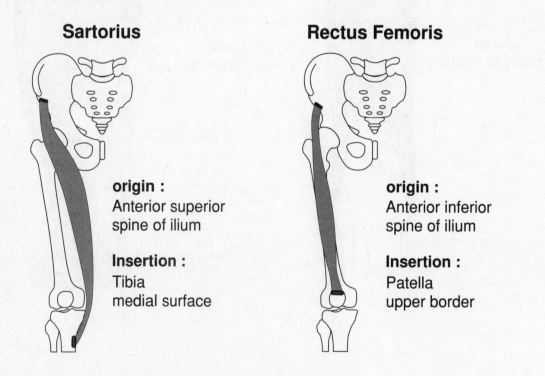

Sartorius

origin :
Anterior superior
spine of ilium

Insertion :
Tibia
medial surface

Rectus Femoris

origin :
Anterior inferior
spine of ilium

Insertion :
Patella
upper border

THIGH : medial muscles

These muscles adduct and flex the thigh.
A pulled groin is the stretching or tearing of these muscles.

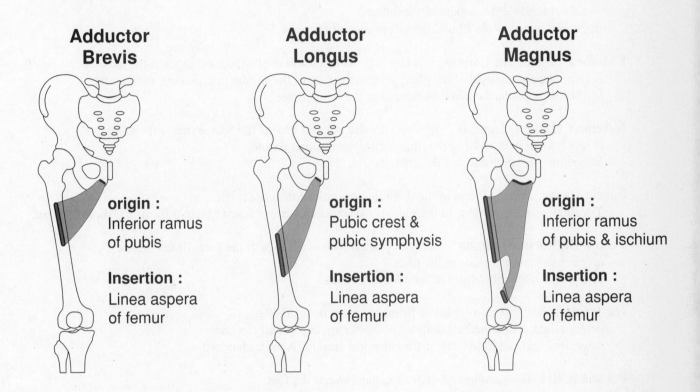

Adductor Brevis

origin :
Inferior ramus
of pubis

Insertion :
Linea aspera
of femur

Adductor Longus

origin :
Pubic crest &
pubic symphysis

Insertion :
Linea aspera
of femur

Adductor Magnus

origin :
Inferior ramus
of pubis & ischium

Insertion :
Linea aspera
of femur

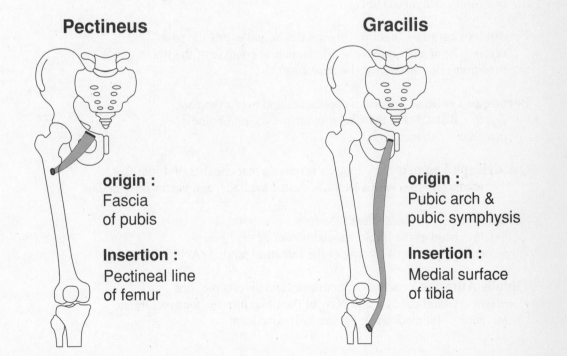

Pectineus

origin :
Fascia
of pubis

Insertion :
Pectineal line
of femur

Gracilis

origin :
Pubic arch &
pubic symphysis

Insertion :
Medial surface
of tibia

Biceps Femoris   *action :* flexes the leg; extends the thigh.
  origin :
       long head : ischial tuberosity.
       short head : linea aspera of the femur.
  insertion : head of the fibula; lateral condyle of the tibia.

Extensor Digitorum Longus   *action :* dorsiflexes and everts the foot; extends the toes.
  origin : lateral condyle of the tibia; anterior surface of the fibula; interosseous membrane.
  insertion : middle and distal phalanges of the 4 outer toes.

Extensor Hallucis Longus   *action :* dorsiflexes and inverts the foot; extends the great toe.
  origin : anterior surface of the fibula; interosseous membrane.
  insertion : distal phalanx of the great toe.

Fascia Lata   A deep fascia of the thigh that encircles the entire thigh.
  Together with the tendons of the gluteus maximus and tensor fasciae latae it forms the iliotibial tract.

Flexor Digitorum Longus   *action :* plantar flexes and inverts the foot; flexes the toes.
  origin : posterior surface of the tibia.
  insertion : distal phalanges of the four outer toes.

Gastrocnemius   *action :* plantar flexes the foot; flexes the leg.
  origin : lateral and medial condyles of the femur; capsule of the knee.
  insertion : calcaneus by way of the calcaneal tendon (Achilles tendon).

Peroneus Brevis   *action :* plantar flexes and everts the foot.
  origin : body of the fibula.
  insertion : 5th metatarsal.

Peroneus Longus   *action :* plantar flexes and everts the foot.
  origin : head and body of the fibula; lateral condyle of the tibia.
  insertion : 1st metatarsal; 1st cuneiform.

Peroneus Tertius   *action :* dorsiflexes and everts the foot.
  origin : distal 3rd of the fibula; interosseous membrane.
  insertion : 5th metatarsal.

Quadriceps Femoris   A composite muscle that consists of 4 muscles :
       rectus femoris, vastus lateralis, vastus medialis, and vastus intermedius.

Soleus   *action :* plantar flexes the foot.
  origin : head of the fibula; medial border of the tibia.
  insertion : calcaneus by way of the calcaneal tendon (Achilles tendon).

Tibialis Anterior   *action :* dorsiflexes and inverts the foot.
  origin : lateral condyle and body of the tibia; interosseous membrane.
  insertion : 1st metatarsal; 1st (medial) cuneiform.

LEG (anterior view)

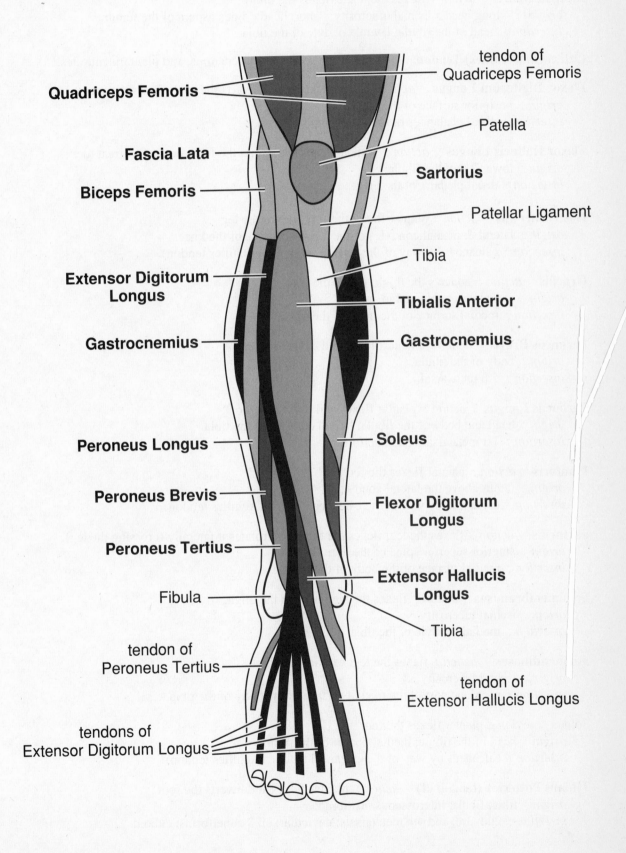

tendon of
Quadriceps Femoris

Quadriceps Femoris

Patella

Fascia Lata

Sartorius

Biceps Femoris

Patellar Ligament

Tibia

Extensor Digitorum
Longus

Tibialis Anterior

Gastrocnemius

Gastrocnemius

Peroneus Longus

Soleus

Peroneus Brevis

Flexor Digitorum
Longus

Peroneus Tertius

Extensor Hallucis
Longus

Fibula

Tibia

tendon of
Peroneus Tertius

tendon of
Extensor Hallucis Longus

tendons of
Extensor Digitorum Longus

SKELETAL MUSCLES / Leg (posterior view)

Biceps Femoris   *action* : flexes the leg; extends the thigh.
  origins :   long head : ischial tuberosity;   short head : linea aspera of the femur.
  insertion : head of the fibula; lateral condyle of the tibia.

Calcaneal (Achilles) Tendon   Tendon of the soleus, gastrocnemius, and plantaris muscles.

Flexor Digitorum Longus   *action* : plantar flexes and inverts the foot; flexes the toes.
  origin : posterior surface of the tibia.
  insertion : distal phalanges of the 4 outer toes.

Flexor Hallucis Longus   *action* : plantar flexes and everts the foot; flexes the great toe.
  origin : lower 2/3 of the fibula.
  insertion : distal phalanx of the great toe.

Gastrocnemius   *action* : plantar flexes the foot; flexes the leg.
  origin : lateral & medial condyles of the femur; capsule of the knee.
  insertion : calcaneus by way of the calcaneal tendon (Achilles tendon).

Gracilis   *action* : adducts the thigh; flexes the leg.
  origin : symphysis pubis; pubic arch.
  insertion : medial surface of the body of the tibia.

Peroneus Brevis   *action* : plantar flexes and everts the foot.
  origin : body of the fibula.
  insertion : 5th metatarsal.

Peroneus Longus   *action* : plantar flexes and everts the foot.
  origin : head and body of the fibula; lateral condyle of the tibia.
  insertion : 1st metatarsal; 1st cuneiform.

Plantaris   *action* : plantar flexes the foot.
  origin : femur above the lateral condyle.
  insertion : calcaneus by way of the calcaneal tendon (Achilles tendon).

Sartorius   *action* : flexes the leg; flexes the thigh and rotates it laterally (crossing the leg).
  origin : anterior superior spine of the ilium.
  insertion : medial surface of the body of the tibia.

Semimembranosus   *action* : flexes the leg; extends the thigh.
  origin : ischial tuberosity.
  insertion : medial condyle of the tibia.

Semitendinosus   *action* : flexes the leg; extends the thigh.
  origin : ischial tuberosity.
  insertion : proximal part of the medial surface of the body of the tibia.

Soleus   *action* : plantar flexes the foot.
  origin : head of the fibula; medial border of the tibia.
  insertion : calcaneus by way of the calcaneal tendon (Achilles tendon).

Tibialis Posterior (tendon of)   *action* : plantar flexes and inverts the foot.
  origin : tibia; fibula; interosseous membrane.
  insertion : 2nd, 3rd, and 4th metatarsals; navicular; all 3 cuneiforms; cuboid.

LEG (posterior view)

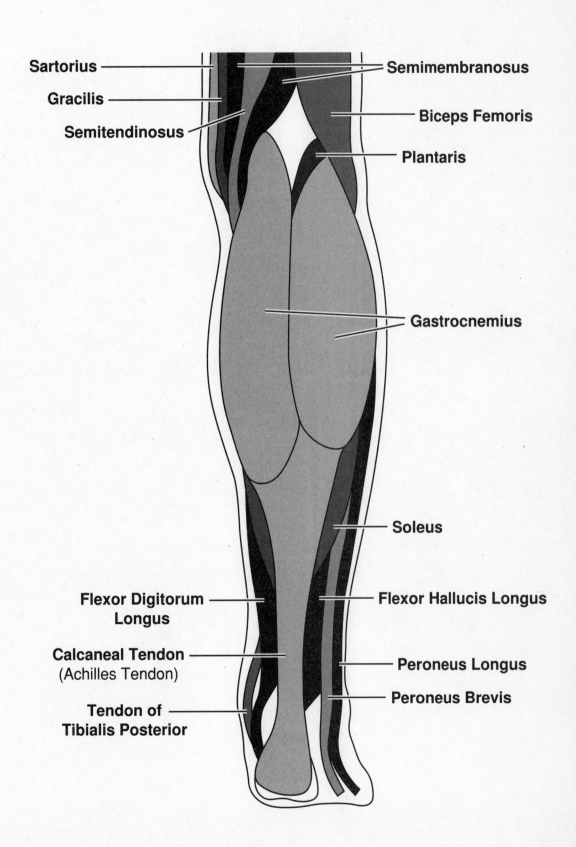

Sartorius

Gracilis

Semitendinosus

Semimembranosus

Biceps Femoris

Plantaris

Gastrocnemius

Soleus

Flexor Digitorum
Longus

Flexor Hallucis Longus

Calcaneal Tendon
(Achilles Tendon)

Peroneus Longus

Peroneus Brevis

Tendon of
Tibialis Posterior

Part II : Self-Testing Exercises

Unlabeled illustrations from Part I

SKELETON : Overview

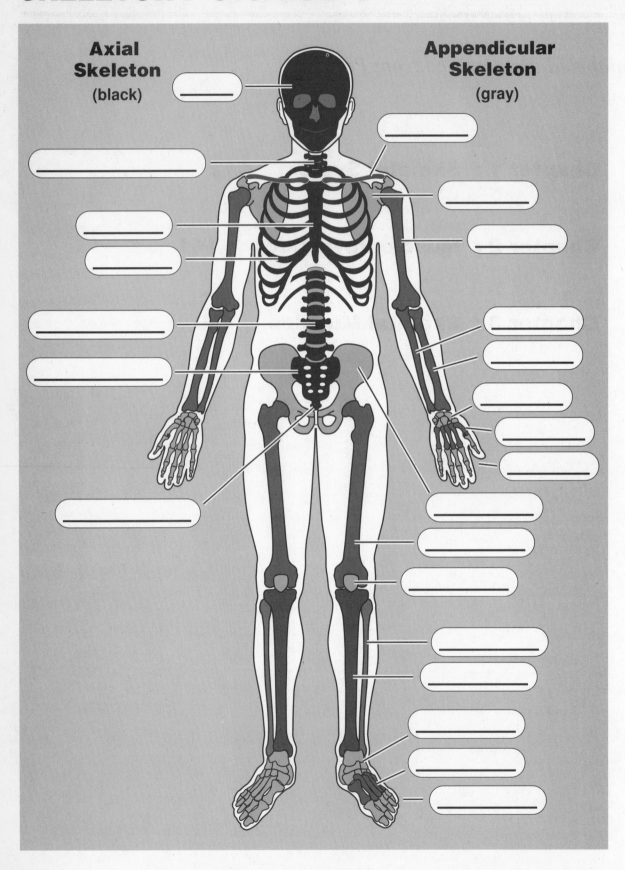

Axial Skeleton (black)

Appendicular Skeleton (gray)

LONG BONE STRUCTURES
Tibia (longitudinal section)

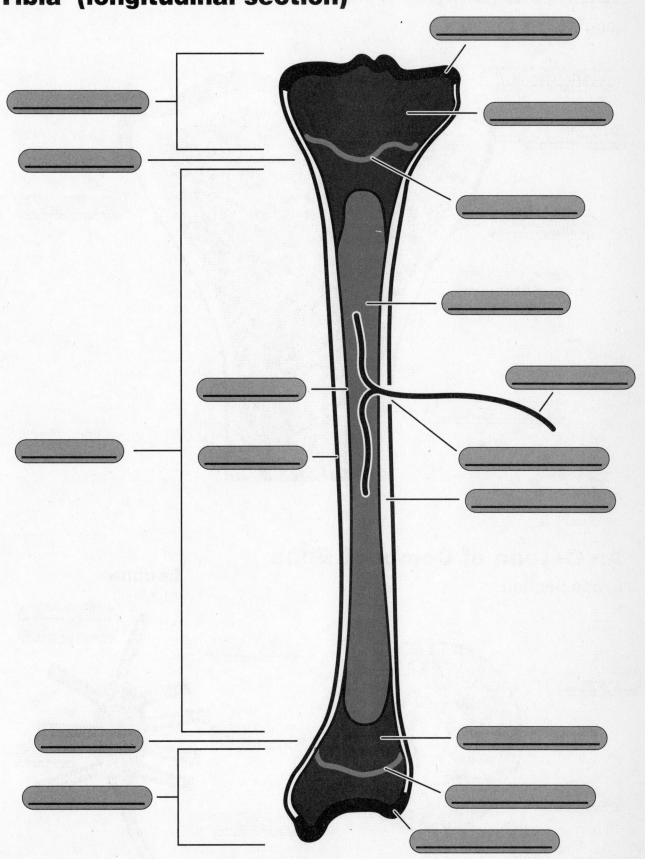

BONE TISSUE

Epiphysis (end) of a Long Bone
Longitudinal Section

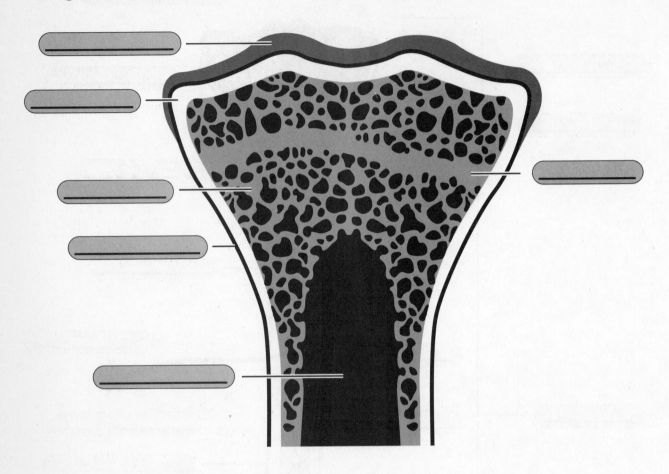

An Osteon of Compact Bone
Cross Section

Lacuna
(detail)

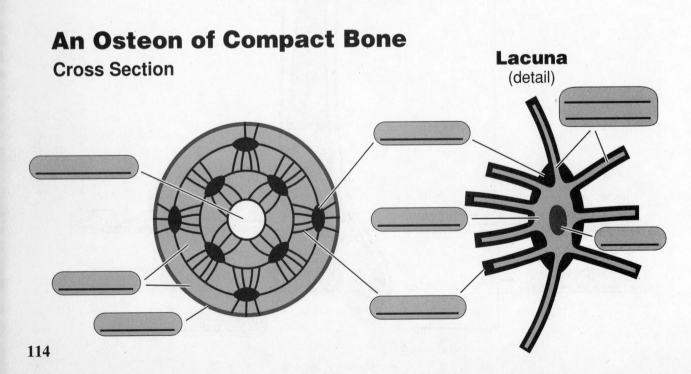

BONE TISSUE : Cell Types and Spongy Bone

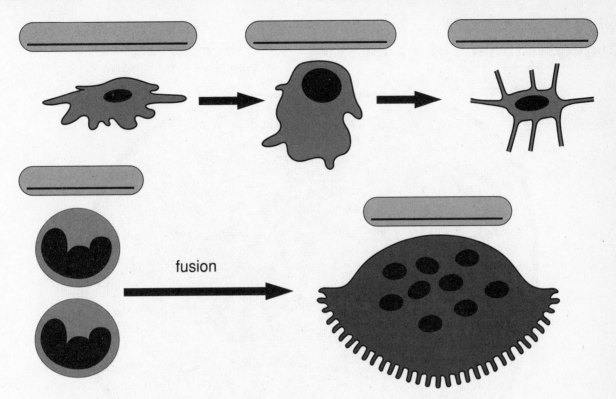

fusion

Spongy Bone Trabeculae

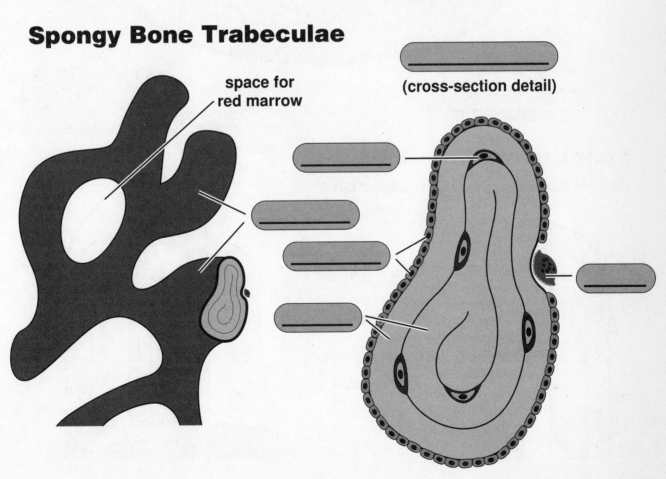

space for
red marrow

(cross-section detail)

BONE TISSUE : Compact Bone

Cross Section
Diaphysis (Shaft) of a Long bone

(contains yellow marrow)

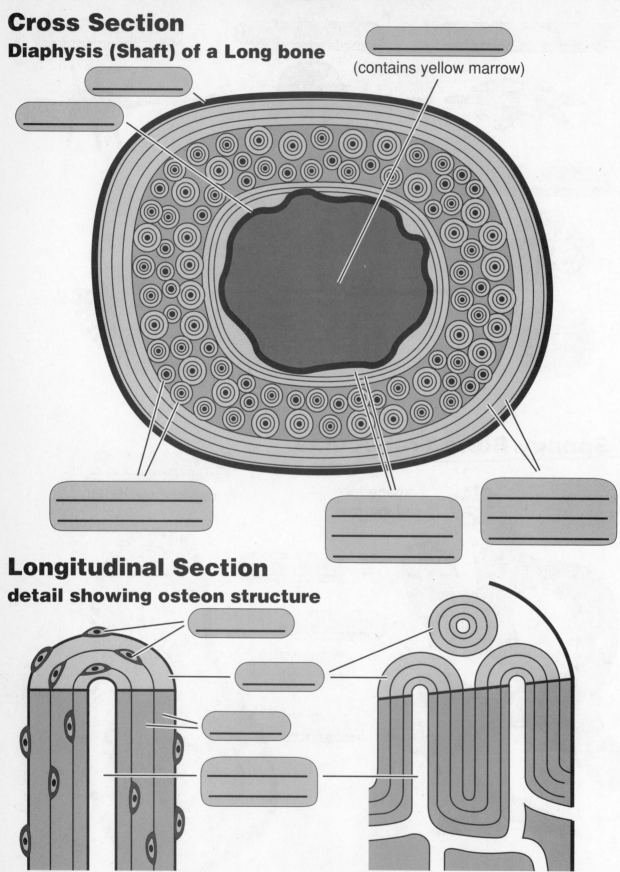

Longitudinal Section
detail showing osteon structure

LONG BONE DEVELOPMENT

1 Cartilaginous Model

2 Periosteal Bone Collar

3 Calcification of Cartilage

4 Primary Ossification Center

5 Continuous Bone Layer

6 Secondary Ossification Centers

7 Fully Developed Bone

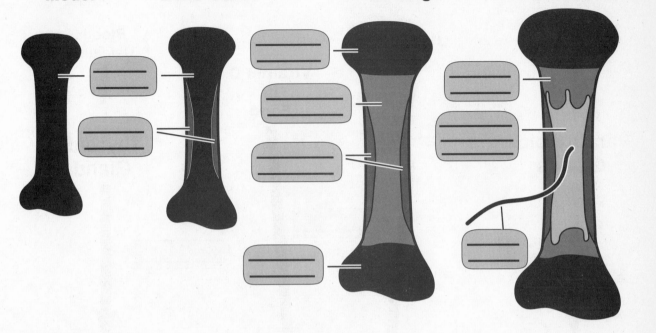
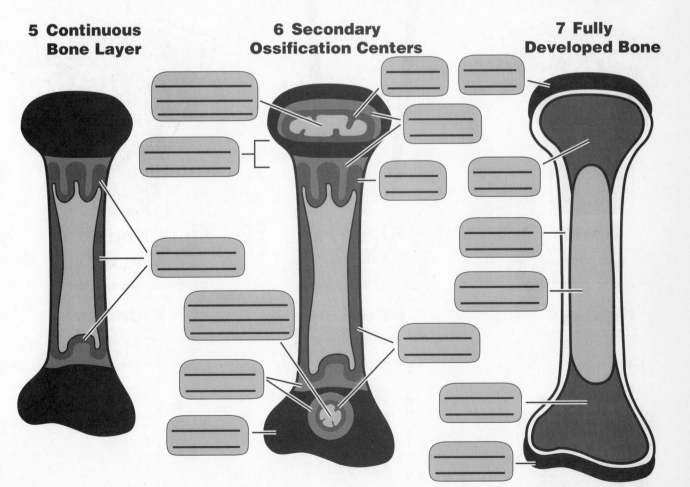

CALCIUM BALANCE

Calcium homeostasis is regulated by Parathyroid Hormone, Vitamin D, and Calcitonin

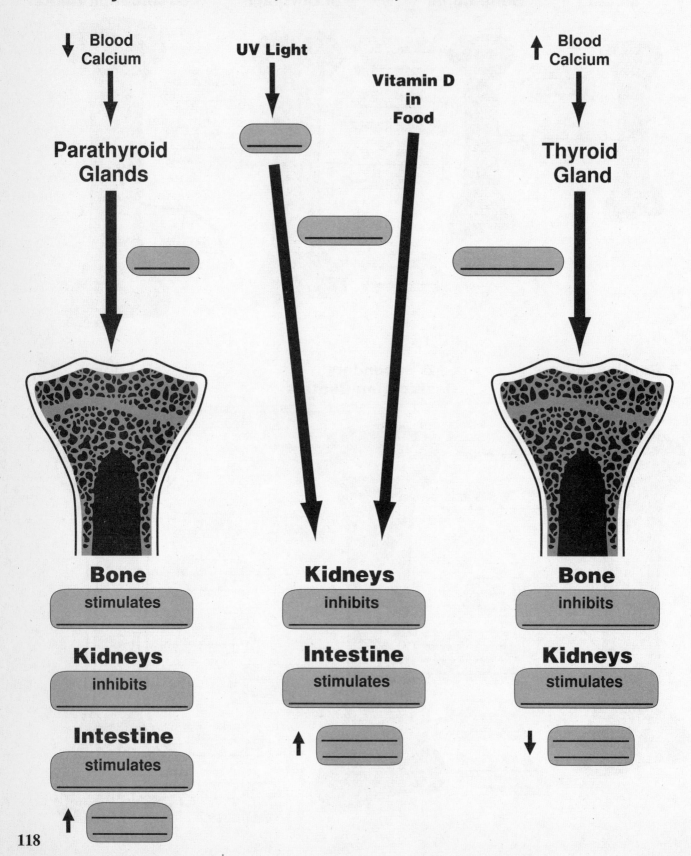

BONE MARKINGS

Upper Extremity
(anterior view)

Lower Extremity
(posterior view)

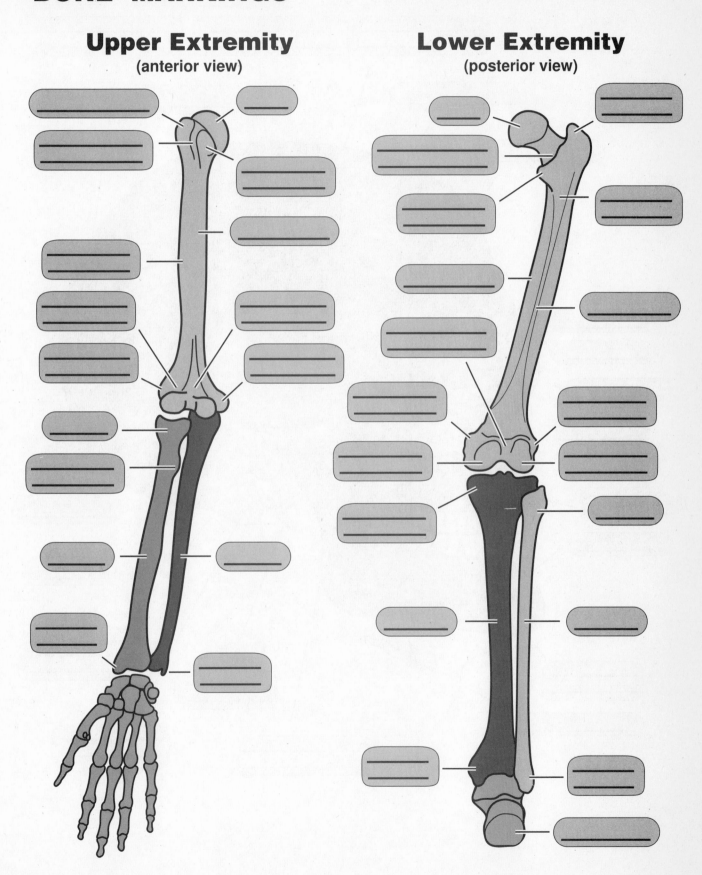

SKULL : lateral view

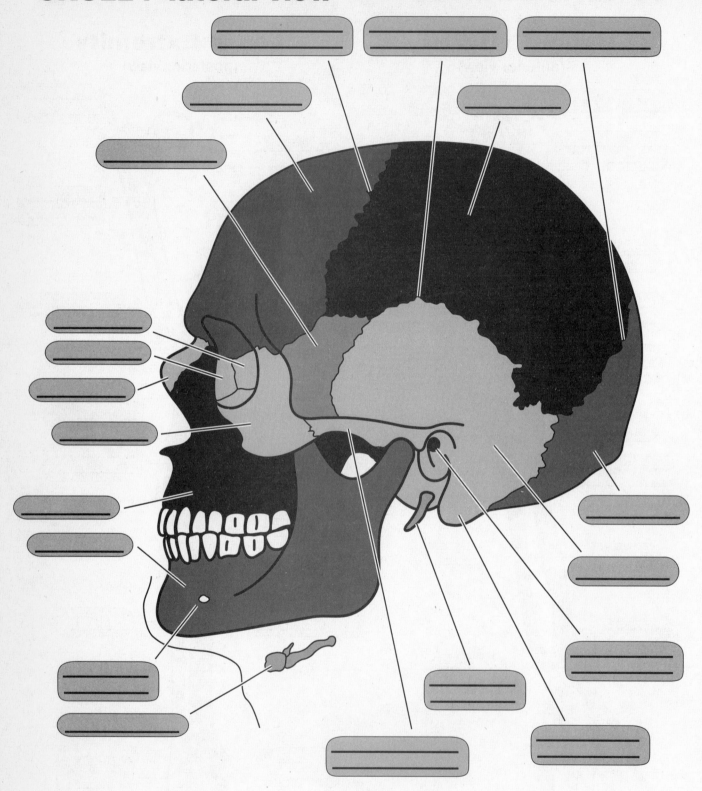

SKULL : anterior view

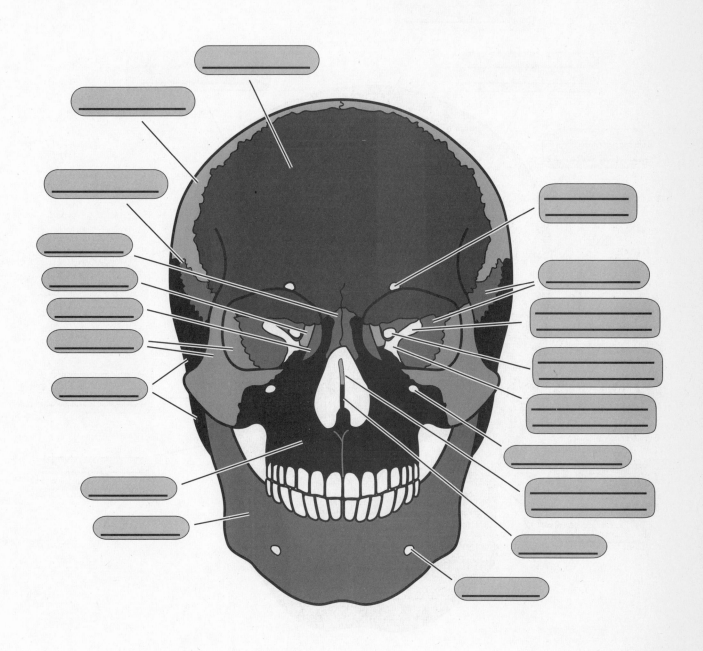

SKULL : posterior view

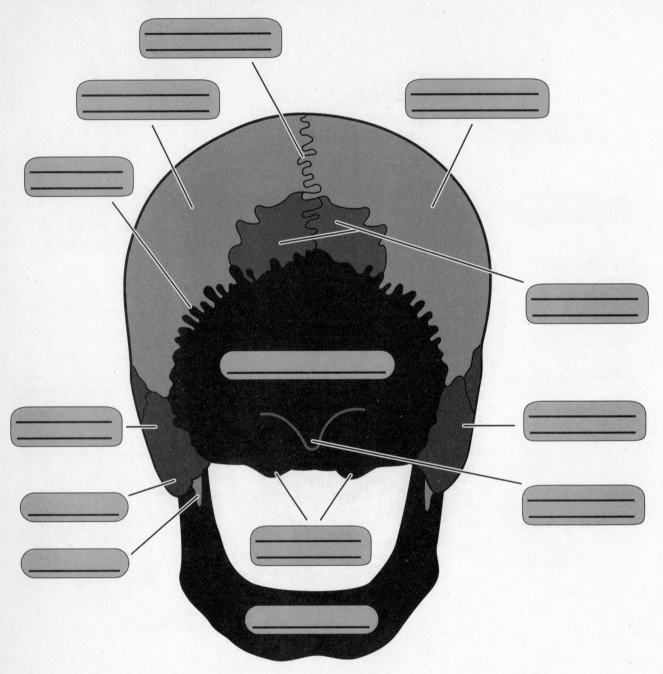

SKULL : floor of the cranium (from above)

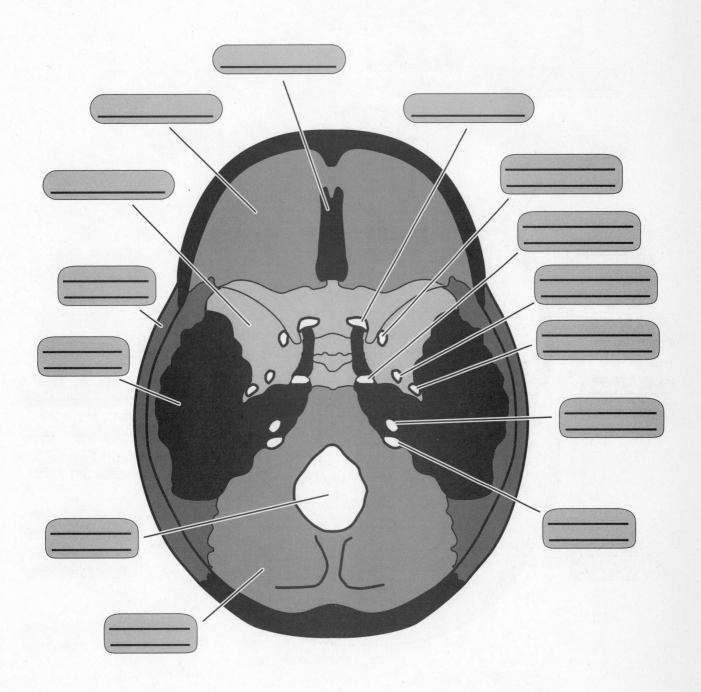

SKULL : inferior view

AUDITORY OSSICLES

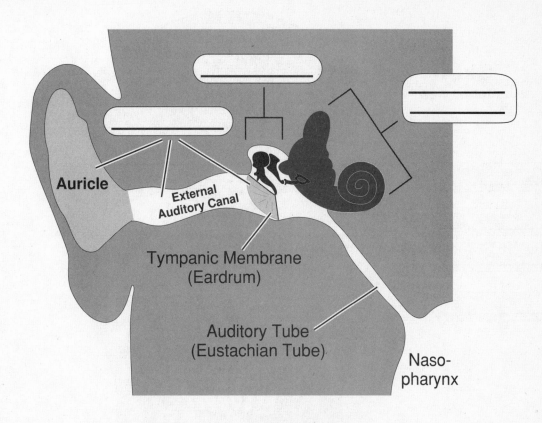

Auricle

External Auditory Canal

Tympanic Membrane
(Eardrum)

Auditory Tube
(Eustachian Tube)

Naso-
pharynx

Middle Ear
(between eardrum and internal ear)

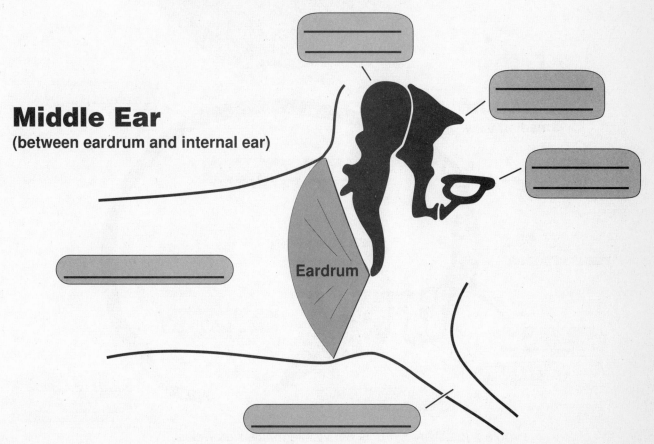

Eardrum

PARANASAL SINUSES

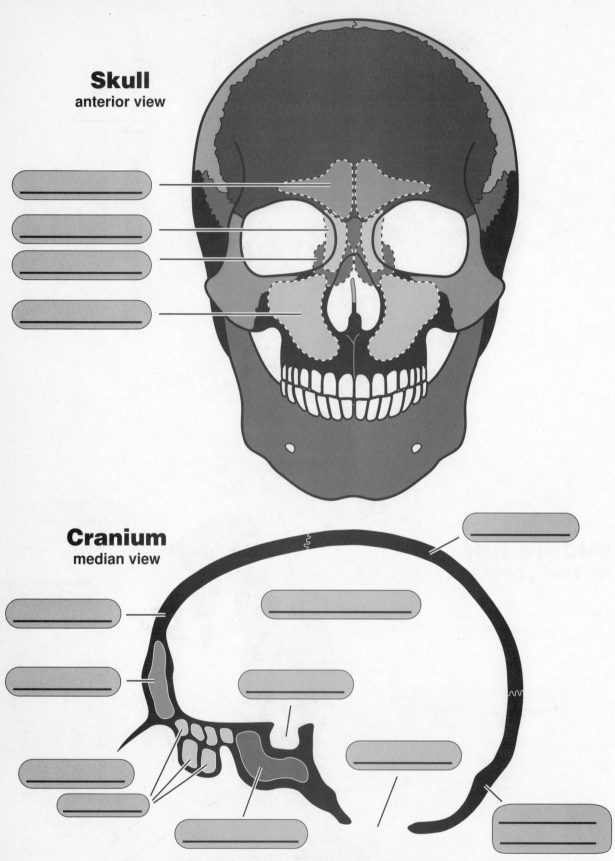

Skull
anterior view

Cranium
median view

note : maxillary sinus is not visible in this illustration

VERTEBRAL COLUMN : 3 Views

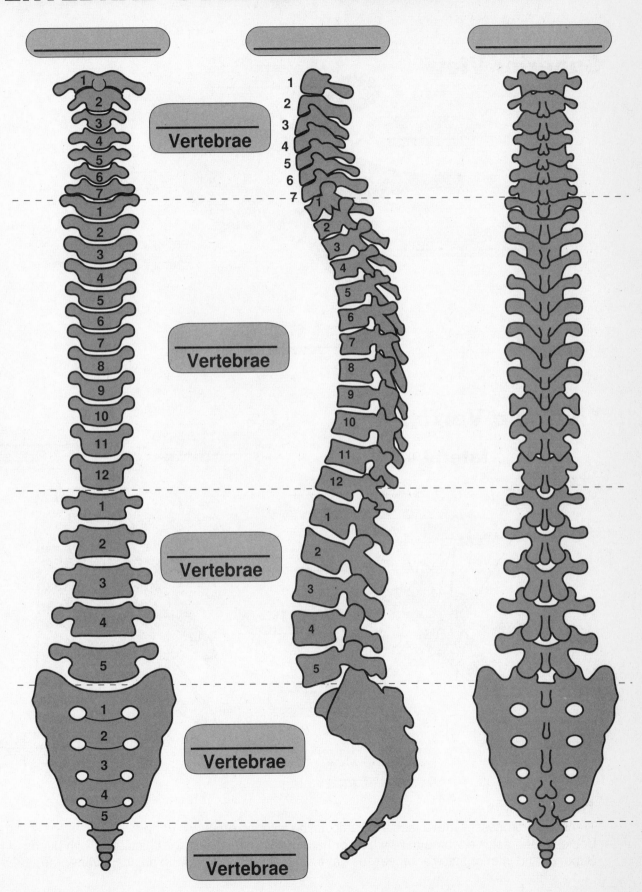

_____

_____

_____

_____ Vertebrae

_____ Vertebrae

_____ Vertebrae

_____ Vertebrae

_____ Vertebrae

TYPICAL VERTEBRA

Superior View

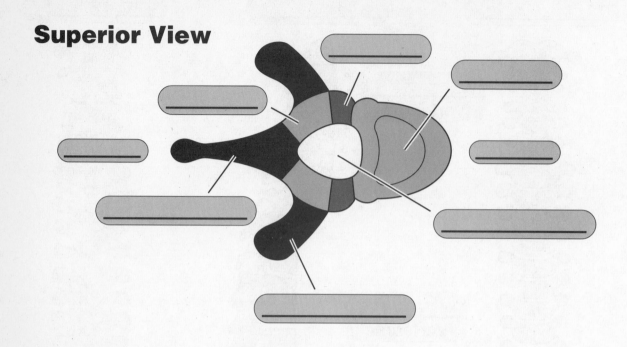

Thoracic Vertebra

lateral view

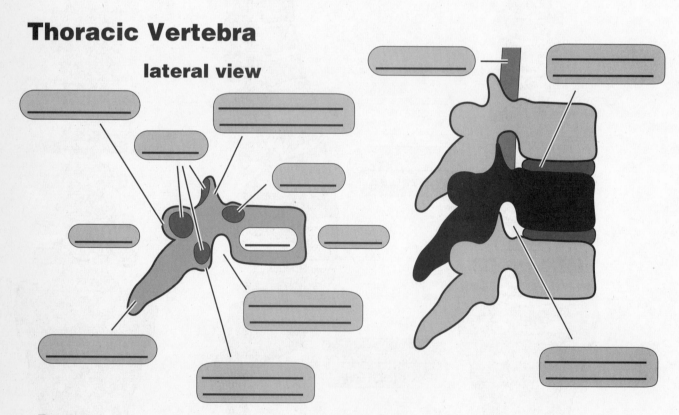

Facets :

Transverse processes have facets for articulating with the tubercles of ribs.
Bodies of thoracic vertebrae have whole or half-facets (demifacets) for articulating with heads of ribs.
Superior and inferior articular processes have facets that articulate with adjacent vertebrae.

VERTEBRAL SHAPES
Superior View

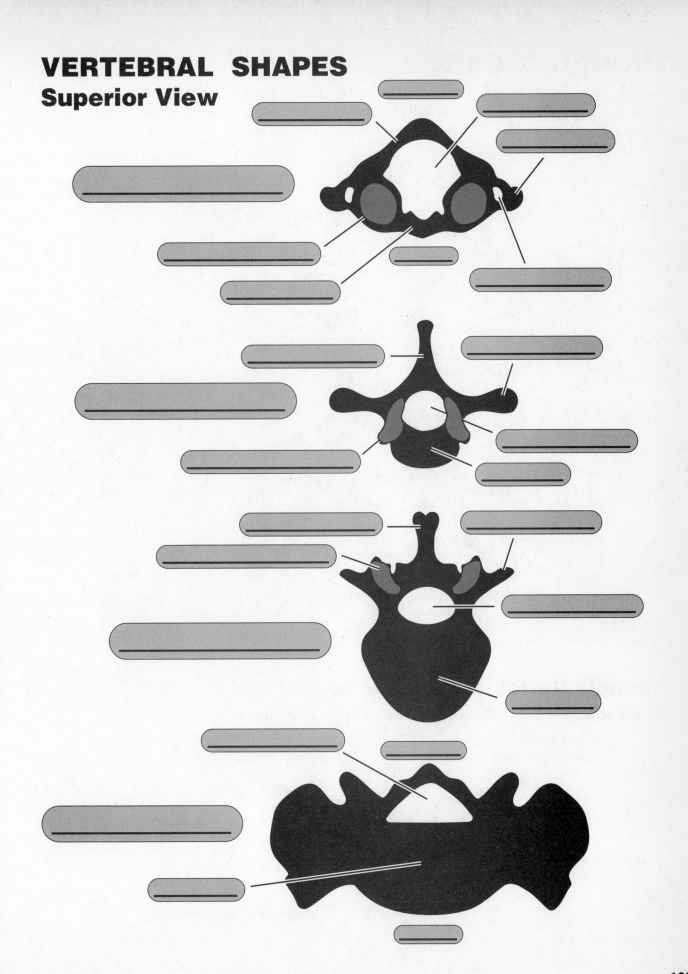

THORACIC CAGE

_____ Ribs (1 — 7) : attach directly to the sternum by costal cartilages.
_____ Ribs (8 — 12) : attach indirectly or not at all to the sternum.
_____ Ribs (11 — 12) : do not attach to the sternum.

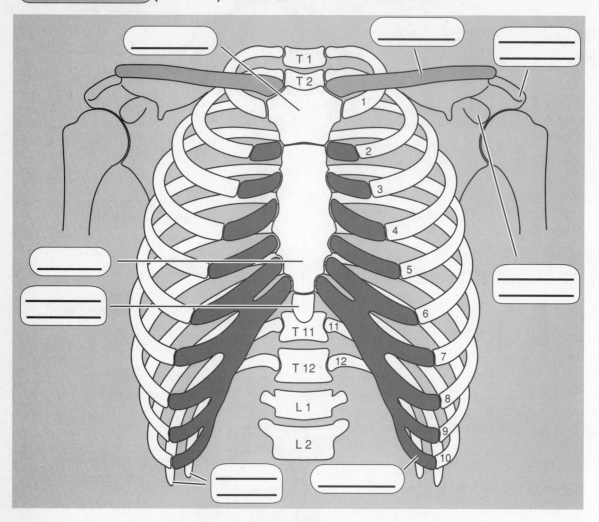

Thoracic Vertebra (viewed from above)

All ribs articulate with their respective thoracic vertebrae.

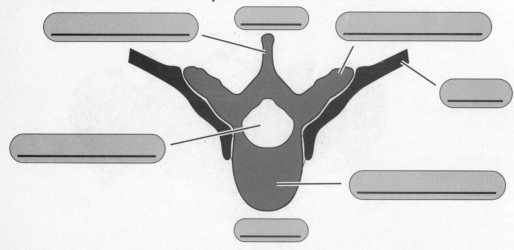

PECTORAL GIRDLE and UPPER EXTREMITY

Anterior View

Posterior View

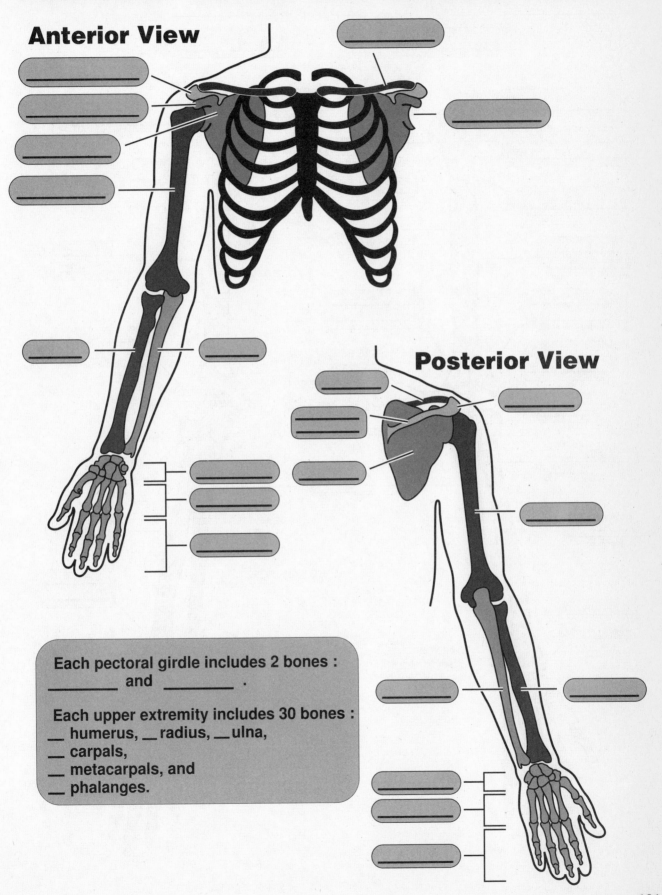

Each pectoral girdle includes 2 bones :
_____ and _____ .

Each upper extremity includes 30 bones :
__ humerus, __ radius, __ ulna,
__ carpals,
__ metacarpals, and
__ phalanges.

UPPER EXTREMITY

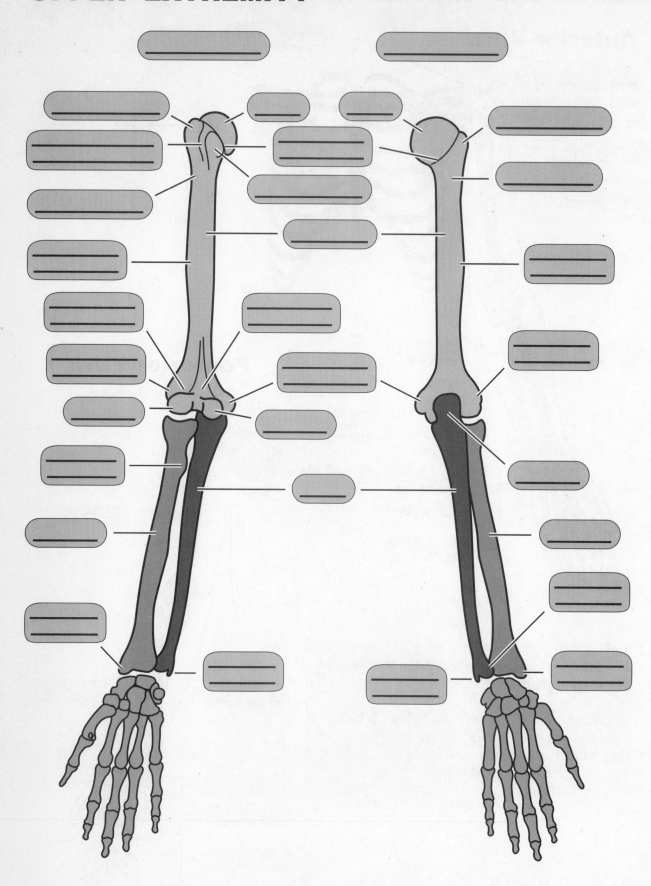

RIGHT HAND

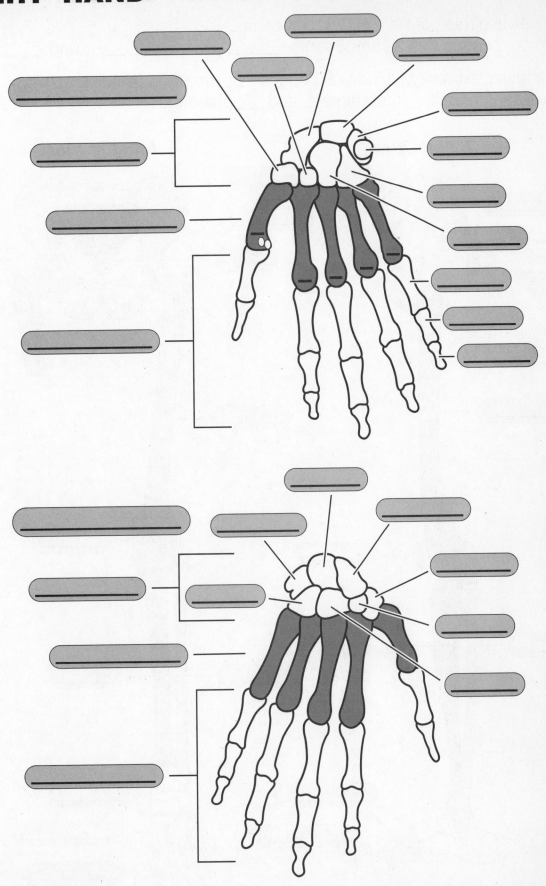

133

PELVIC GIRDLE and LOWER EXTREMITY

The pelvic girdle consists of 2 coxal bones.
Each coxal bone has 3 components : _____ , _____ , and _____ .

Each lower extremity includes 30 bones : __ femur, __ patella, __ tibia,
__ fibula, __ tarsals, __ metatarsals, and __ phalanges.

Lateral View

Anterior View

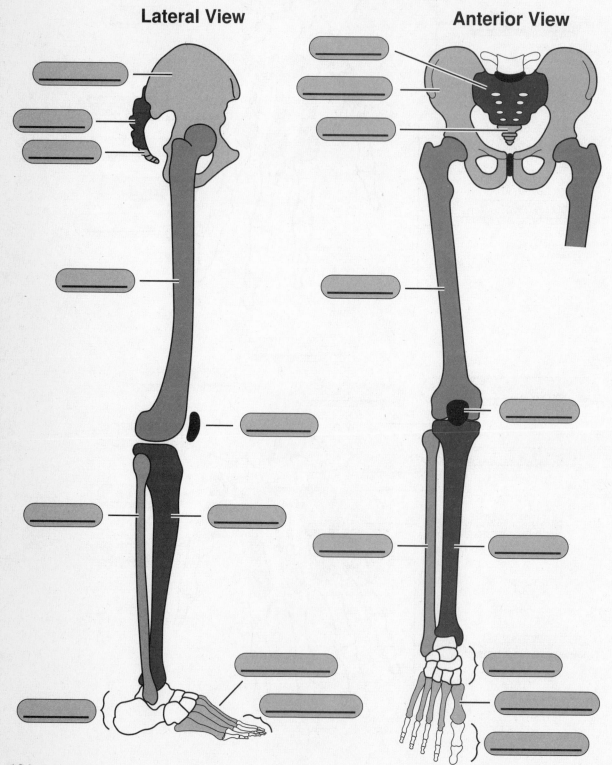

LOWER EXTREMITY

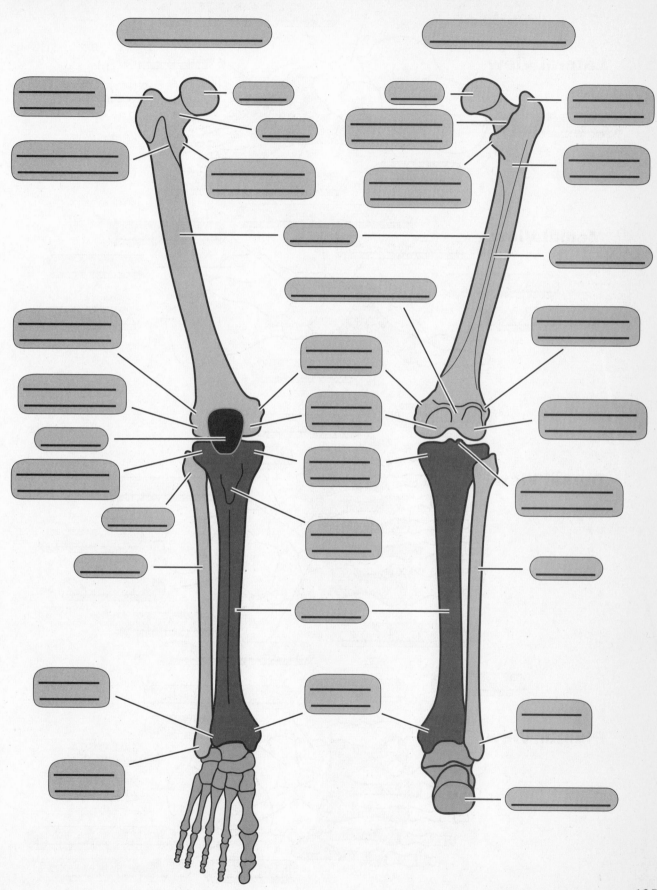

RIGHT FOOT

Lateral View

Medial View

Dorsal View

Plantar View

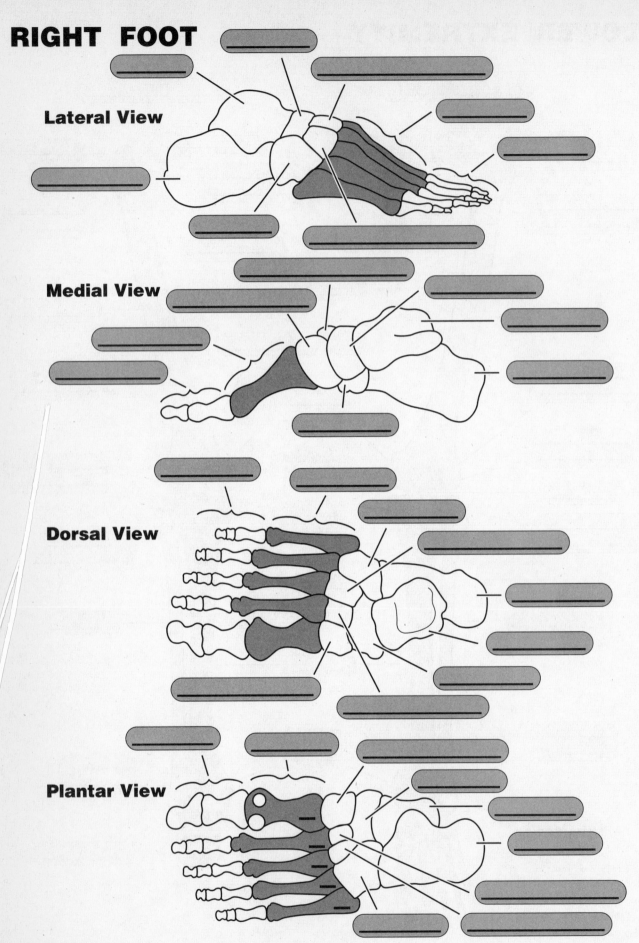

PELVIS

Pelvis : includes the pelvic girdle, sacrum, and coccyx.
Pelvic Girdle : consists of the 2 coxal bones.
Coxal Bone : has 3 fused components (ilium, ischium, and pubis).

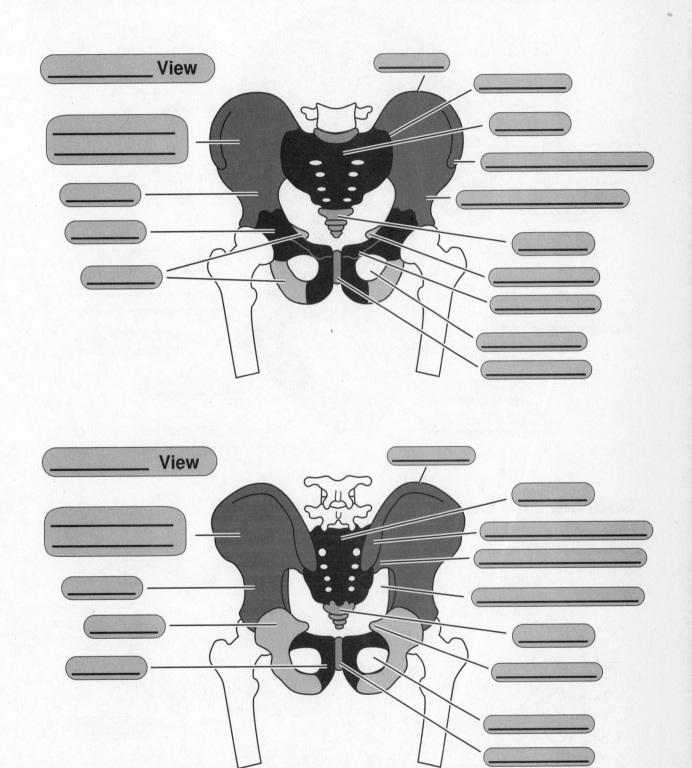

_____ View

_____ View

COXAL BONE and SACRUM

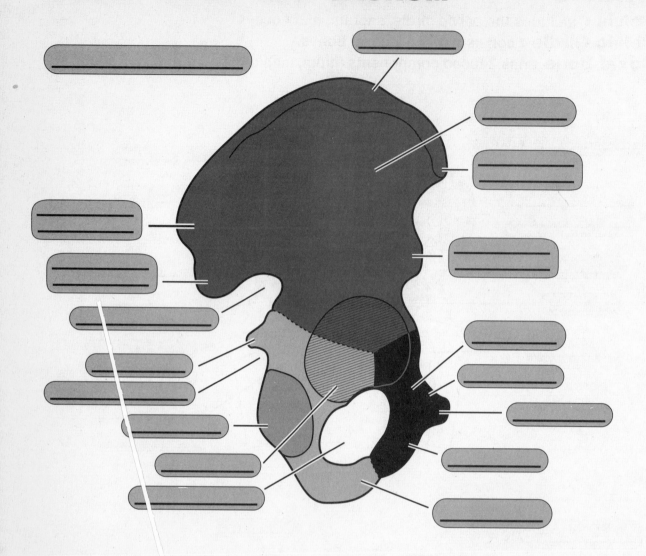

Sacrum and Coccyx

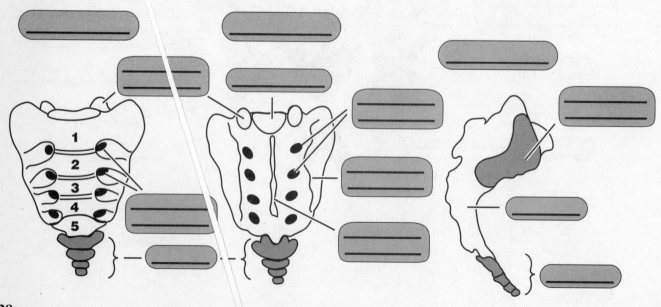

JOINTS (Articulations)

_____
(immovable joint)

_____   _____
_____

_____

_____

_____

_____

_____
(slightly movable joint)

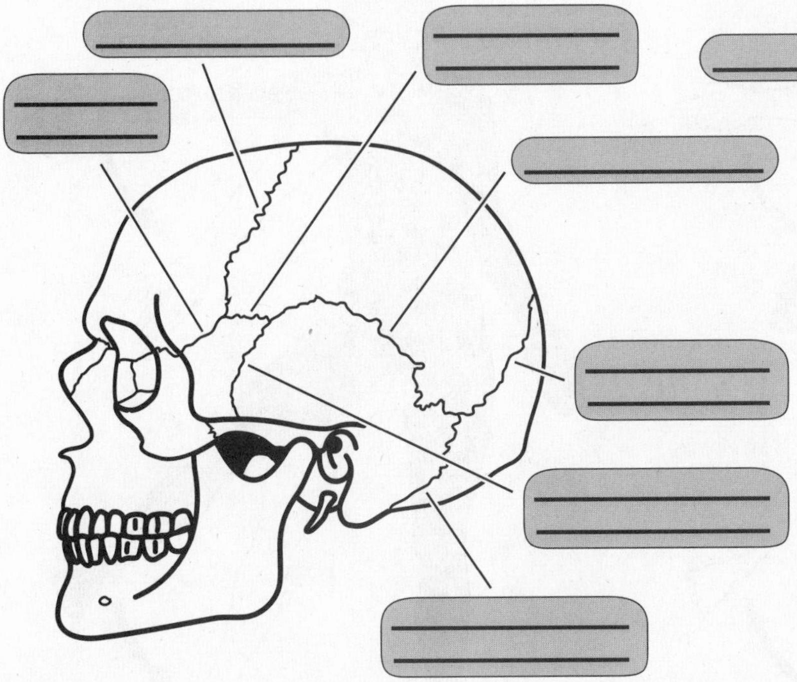

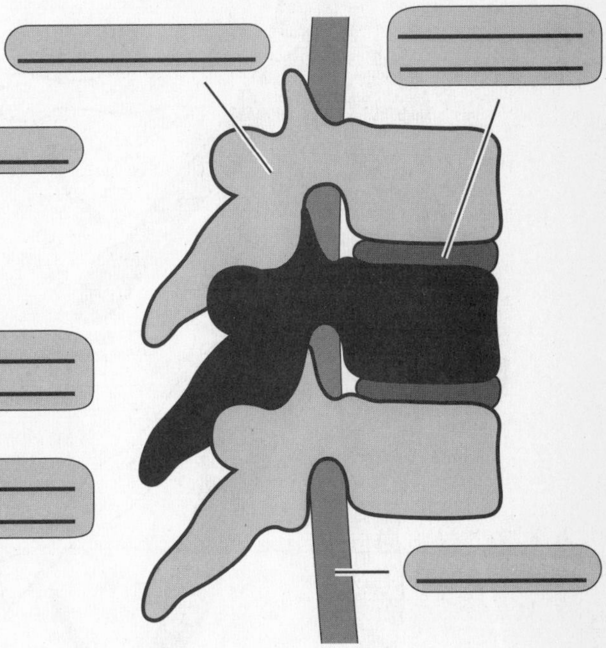

_____

_____

_____

_____
(freely movable joint)

_____
(generalized)

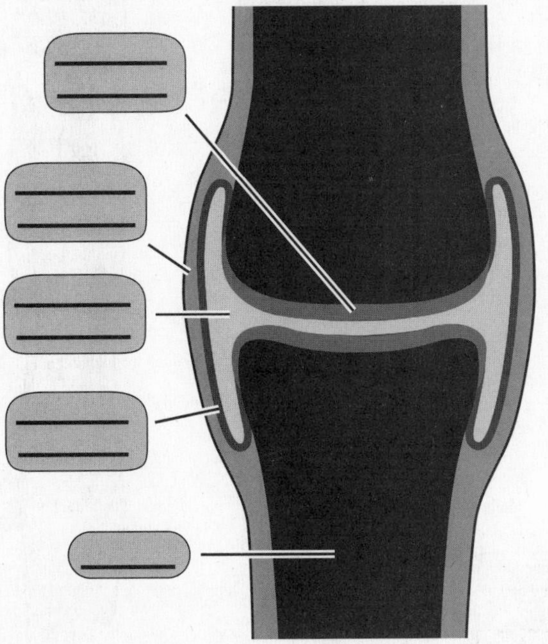

_____

_____

_____

_____

_____

Knee Joint

_____ : Synovial Gliding Type
_____ : Synovial Hinge Type
_____ : Synovial Hinge Type

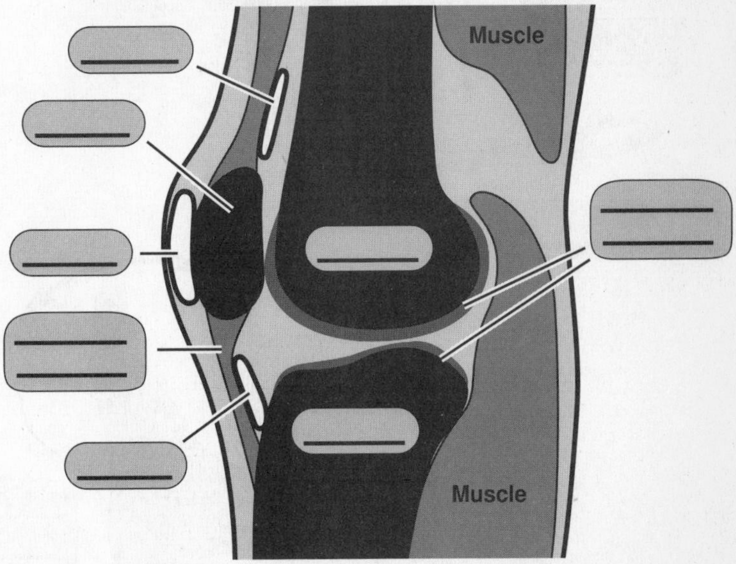

Muscle

_____

_____

_____

_____

Muscle

139

MOVEMENTS AT SYNOVIAL JOINTS

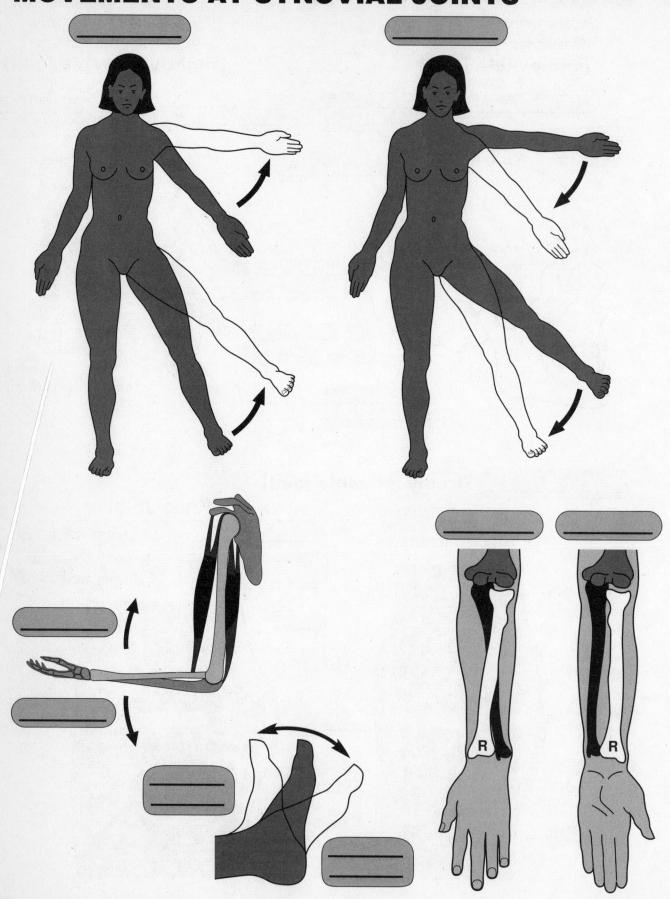

MUSCLE TISSUE

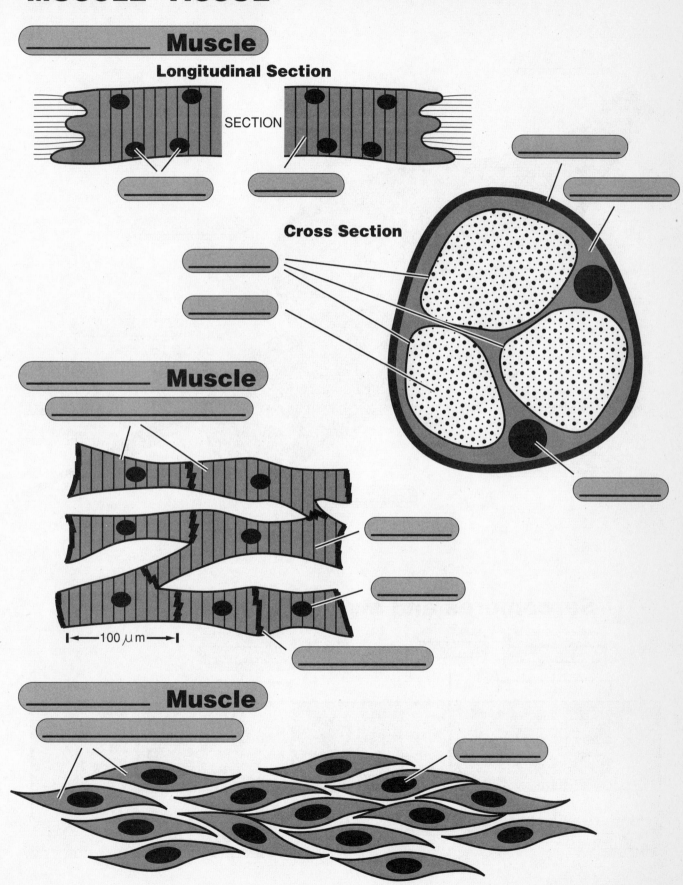

_____ **Muscle**

Longitudinal Section

SECTION

Cross Section

Muscle

I◄—100 μm—►I

Muscle

SKELETAL MUSCLE ANATOMY

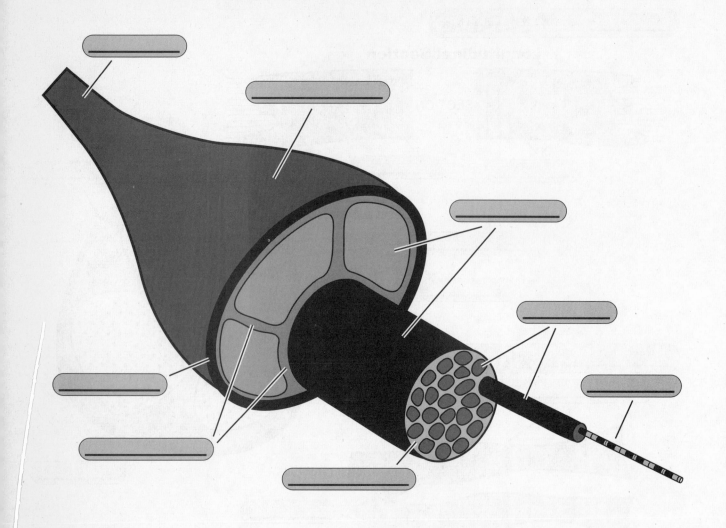

Sarcomeres and Myofibril Bands

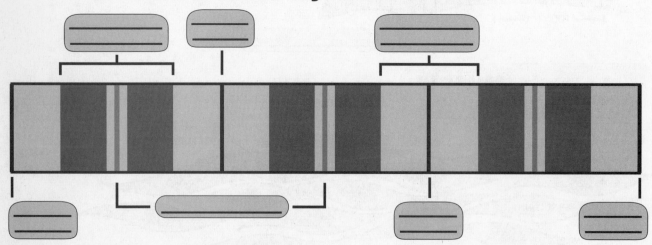

SKELETAL MUSCLE FIBER

Portion of a Muscle Fiber

Myofibril

Sarcomeres

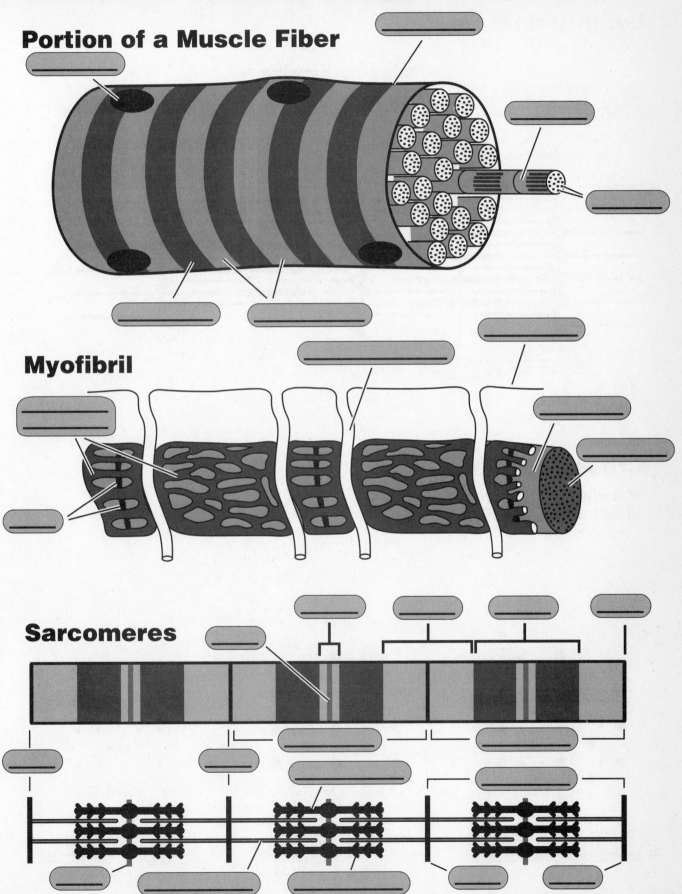

143

MYOFILAMENTS
Longitudinal Section

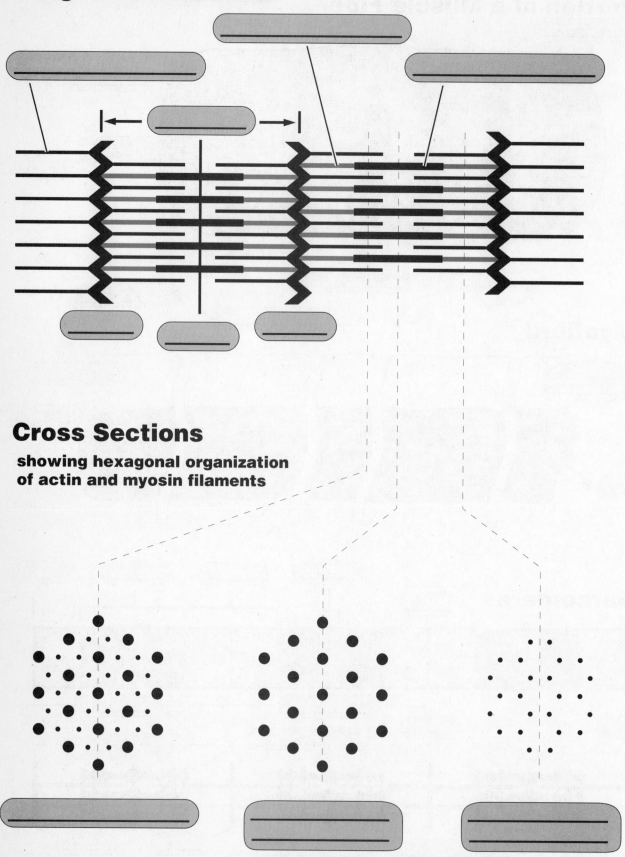

Cross Sections

**showing hexagonal organization
of actin and myosin filaments**

ARRANGEMENT OF FASCICULI

Fasciculi : bundles of skeletal muscle fibers.
Skeletal muscle fibers are arranged in a parallel fashion within each bundle,
but the arrangement of the fasciculi with respect to the tendons may take
several characteristic patterns : parallel, circular, fusiform, and pennate.

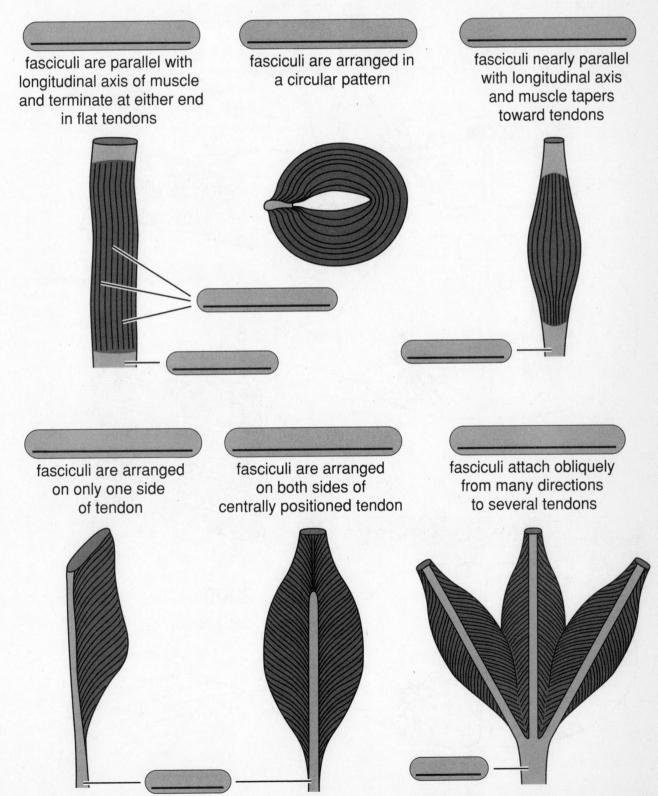

fasciculi are parallel with longitudinal axis of muscle and terminate at either end in flat tendons

fasciculi are arranged in a circular pattern

fasciculi nearly parallel with longitudinal axis and muscle tapers toward tendons

fasciculi are arranged on only one side of tendon

fasciculi are arranged on both sides of centrally positioned tendon

fasciculi attach obliquely from many directions to several tendons

MOTOR UNITS

Motor Unit : the number of muscle fibers innervated by a single motor neuron

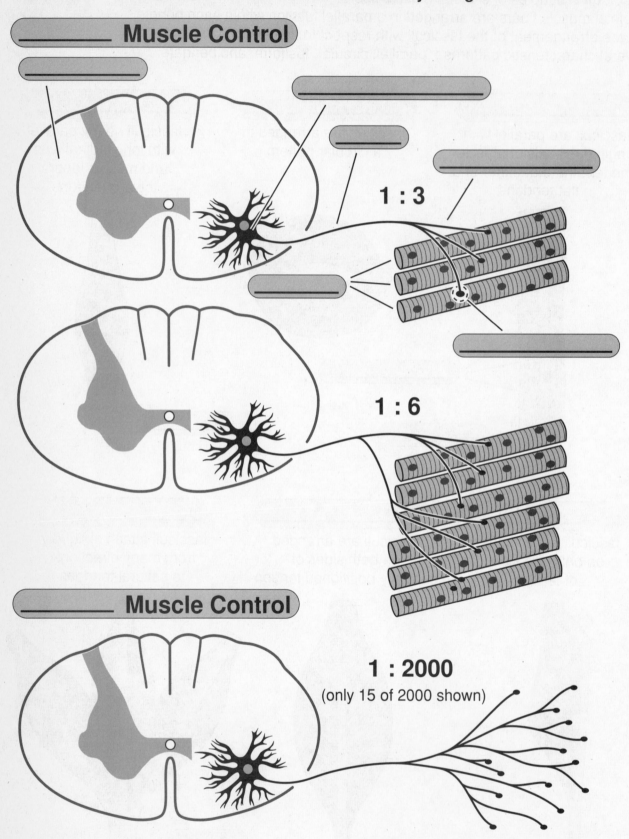

_____ Muscle Control

_____

_____

_____

1 : 3

_____

_____

_____

1 : 6

_____ Muscle Control

1 : 2000

(only 15 of 2000 shown)

NEUROMUSCULAR JUNCTION

Structures

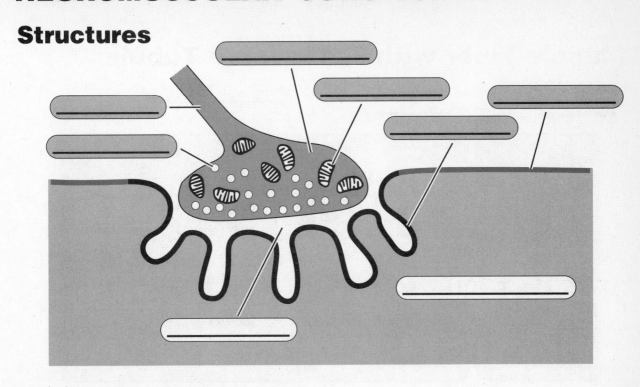

Impulse Transmission

ACh = _____

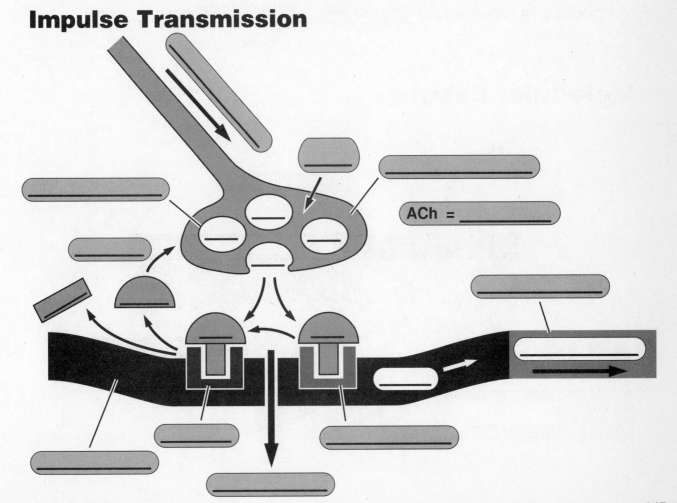

EXCITATION - CONTRACTION COUPLING

Muscle Fiber with Transverse Tubule

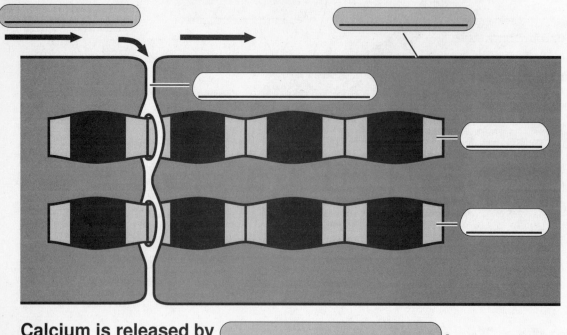

Calcium is released by _____ .

Molecular Events

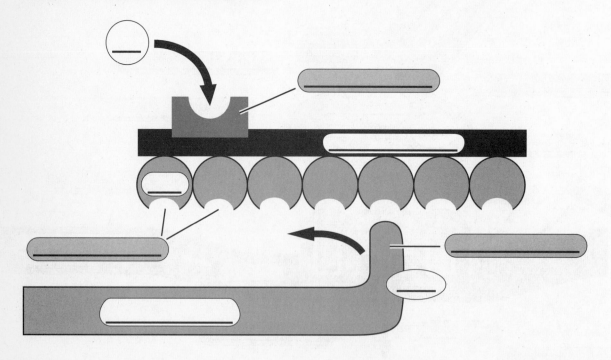

SLIDING FILAMENT MECHANISM

Relaxed Muscle Fiber

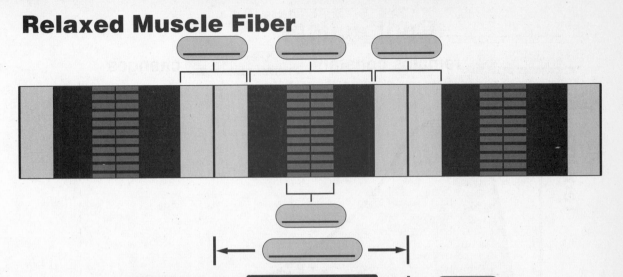

Contracted Muscle Fiber

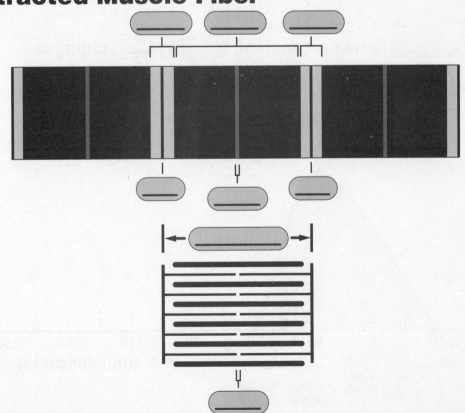

ISOMETRIC VS. ISOTONIC CONTRACTION

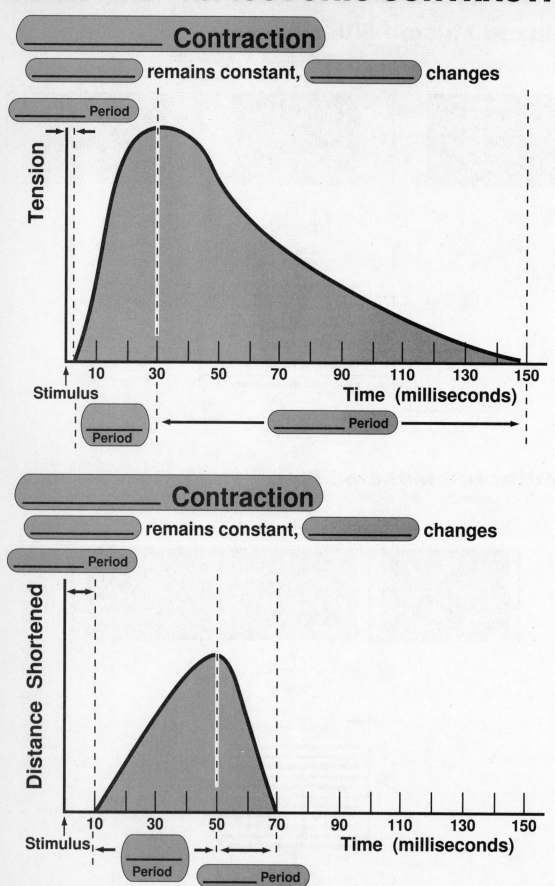

_____ Contraction

_____ remains constant, _____ changes

_____ Period

Tension

Stimulus

Time (milliseconds)

10   30   50   70   90   110   130   150

_____ Period

_____ Period

_____ Contraction

_____ remains constant, _____ changes

_____ Period

Distance Shortened

Stimulus

Time (milliseconds)

10   30   50   70   90   110   130   150

_____ Period

_____ Period

SUMMATION

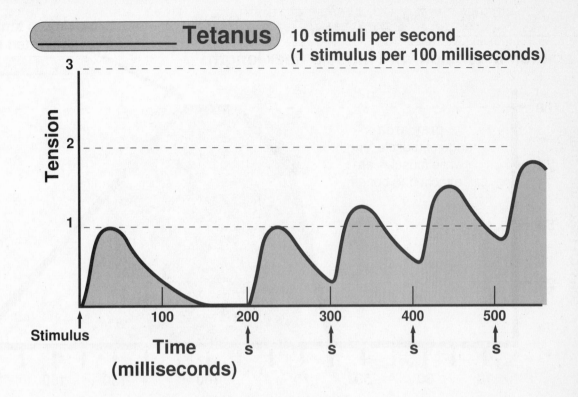

Tetanus — 10 stimuli per second
(1 stimulus per 100 milliseconds)

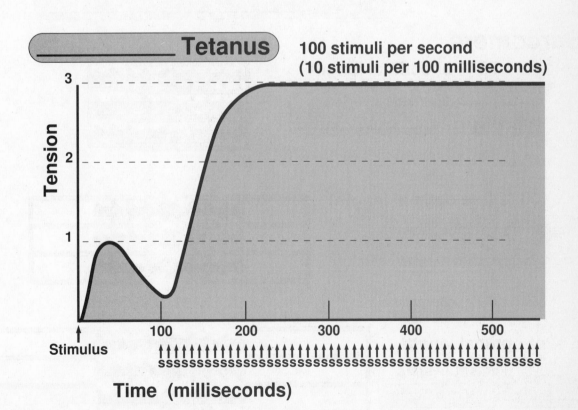

Tetanus — 100 stimuli per second
(10 stimuli per 100 milliseconds)

LENGTH - TENSION RELATIONSHIP

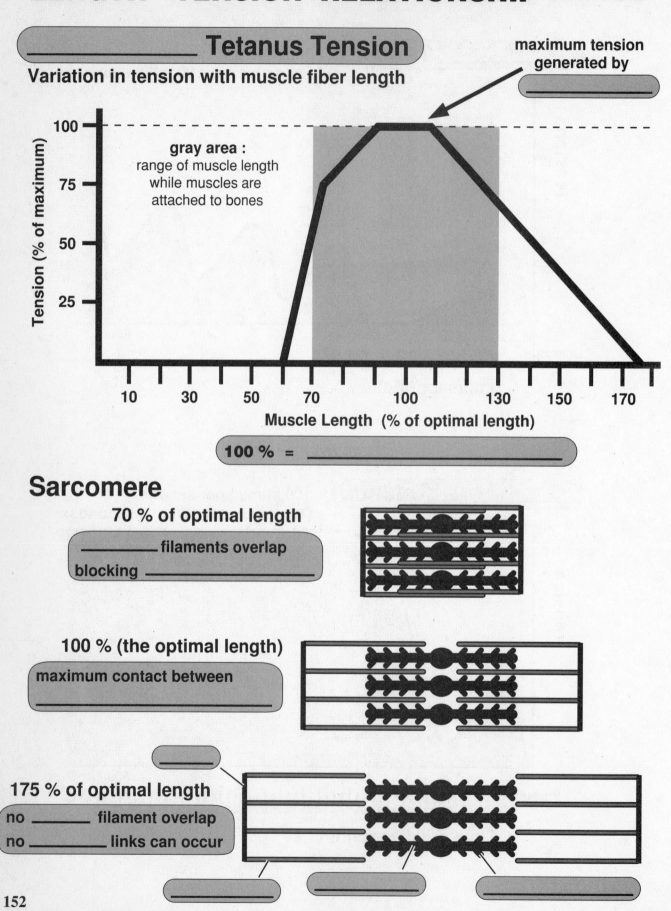

_____ Tetanus Tension

maximum tension generated by
_____

Variation in tension with muscle fiber length

gray area :
range of muscle length
while muscles are
attached to bones

Tension (% of maximum)

100
75
50
25

10   30   50   70   100   130   150   170

Muscle Length (% of optimal length)

100 % = _____

Sarcomere

70 % of optimal length

_____ filaments overlap
blocking _____

100 % (the optimal length)

maximum contact between
_____

175 % of optimal length

_____

no _____ filament overlap
no _____ links can occur

_____   _____   _____

152

ENERGY SOURCES

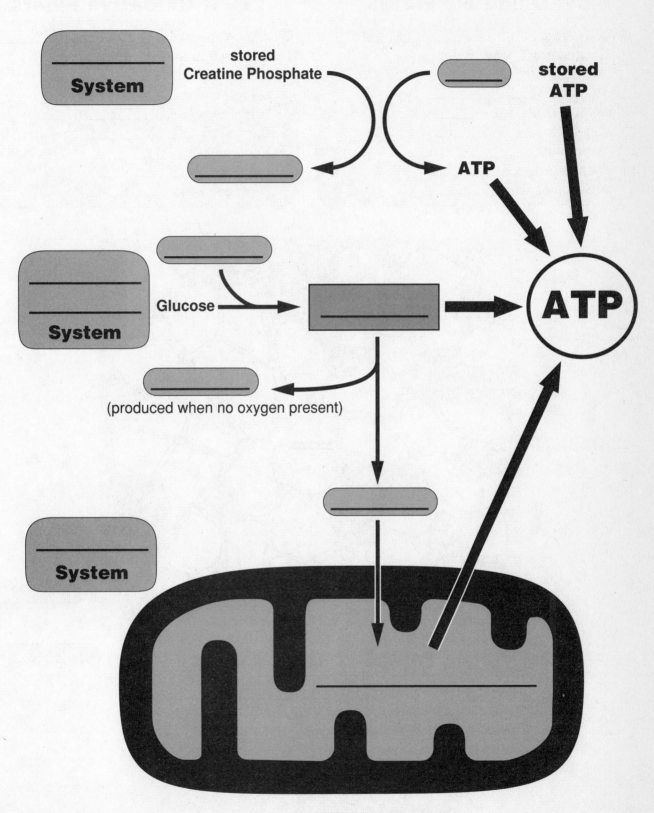

Mitochondrion

SKELETAL MUSCLE FIBER TYPES
Cross Section

Slow Oxidative Fibers

ATP Production : _____
Contraction Speed : _____
ATPase Activity : _____
Myoglobin : _____
Mitochondria : _____
Capillaries : _____
Endurance : _____
Diameter : _____
Fiber color : _____

Fast Oxidative Fibers

ATP Production : _____
Contraction Speed : _____
ATPase Activity : _____
Myoglobin : _____
Mitochondria : _____
Capillaries : _____
Endurance : _____
Diameter : _____
Fiber color : _____

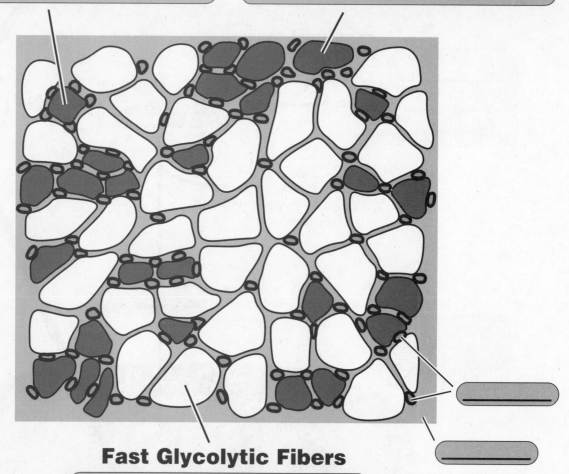

Fast Glycolytic Fibers

ATP Production : _____
Contraction Speed : _____
ATPase Activity : _____
Myoglobin : _____
Mitochondria : _____
Capillaries : _____
Endurance : _____
Diameter : _____
Fiber color : _____

154

TENDON ORGAN and MUSCLE SPINDLE

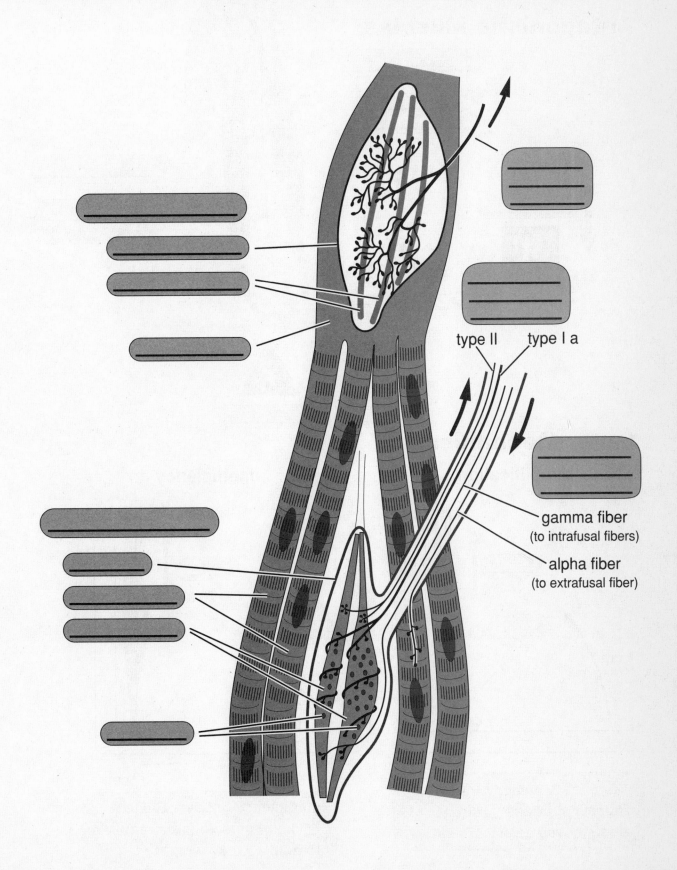

type II   type I a

gamma fiber
(to intrafusal fibers)

alpha fiber
(to extrafusal fiber)

LEVER SYSTEMS

Antagonistic Muscles

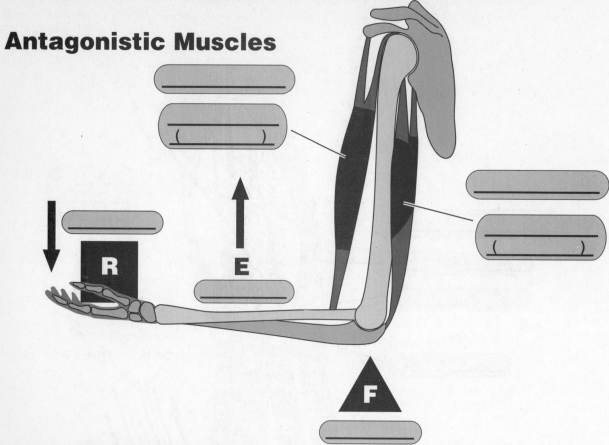

R

E

F

Lever Mechanisms

Amplification

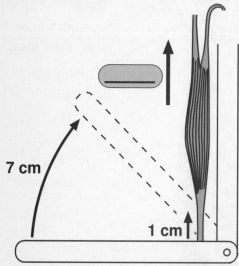

7 cm

1 cm

Distance & Velocity Amplified :
muscle shortens ___ cm
hand moves ___ cm

Inefficiency

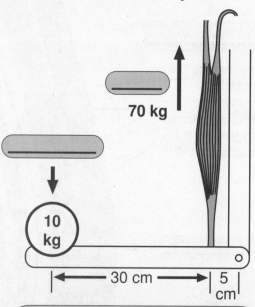

70 kg

10 kg

30 cm

5 cm

Inefficiency of Lever System :
___ kg of muscle force required
to hold a ___ kg weight

SMOOTH MUSCLE

Single-Unit

Fibers connected by _____ form networks.
A few fibers are directly stimulated by neurons;
action potentials spread via _____ to many adjacent fibers.

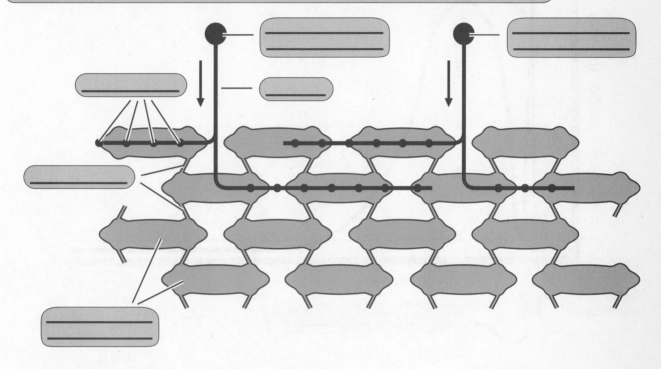

Multi-Unit

Multiunit fibers act _____ .
Each fiber is directly stimulated by _____ located on branching **axons**.

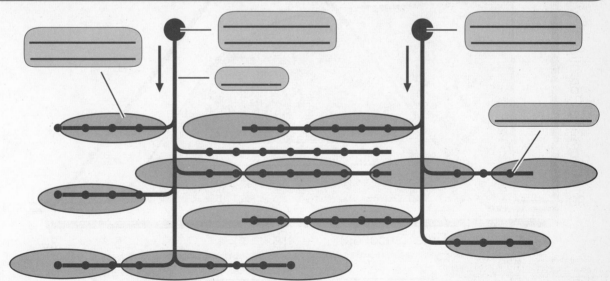

REFRACTORY PERIOD

The refractory period lasts as long as the ⬭ .
During this time the muscle membrane cannot respond to stimuli.

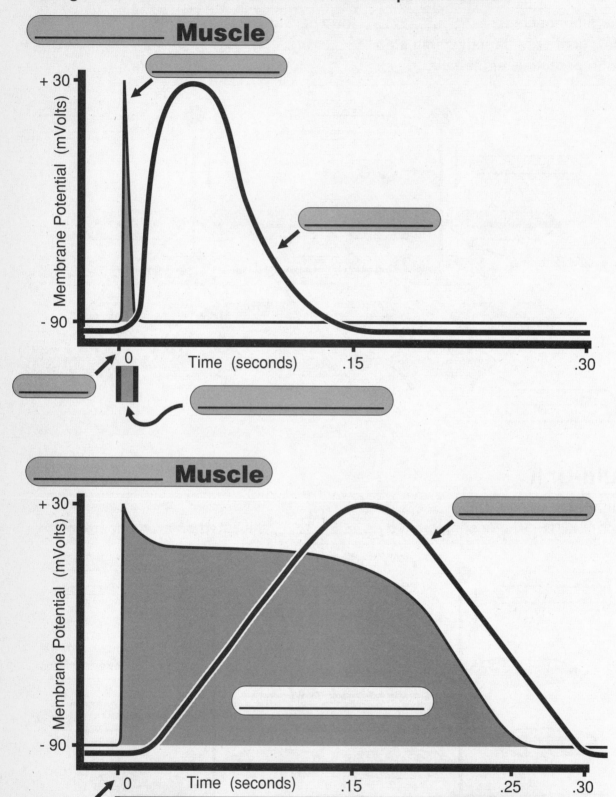

_____ **Muscle**

_____ **Muscle**

MAJOR MUSCLES : Anterior View

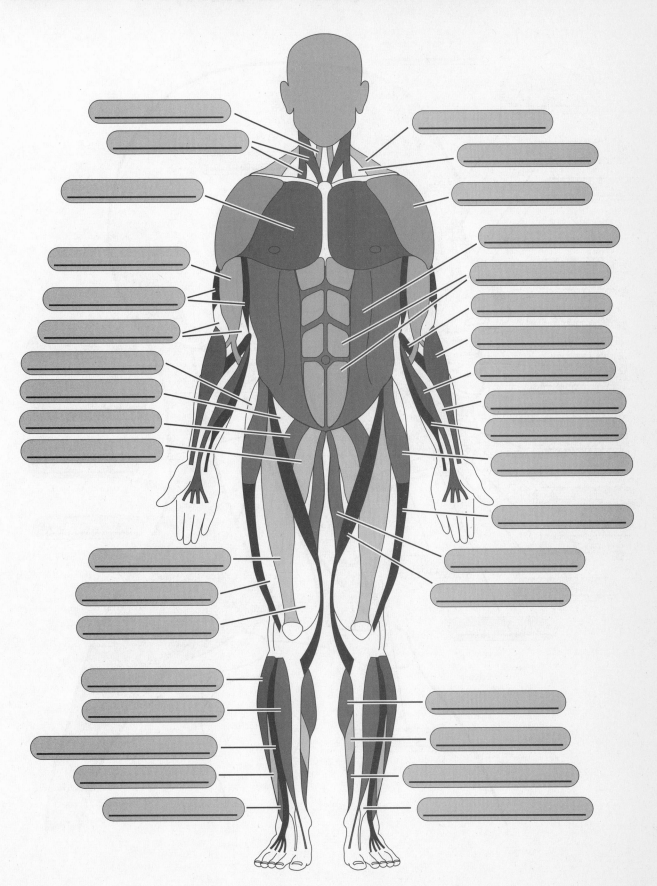

HEAD and NECK (lateral view)

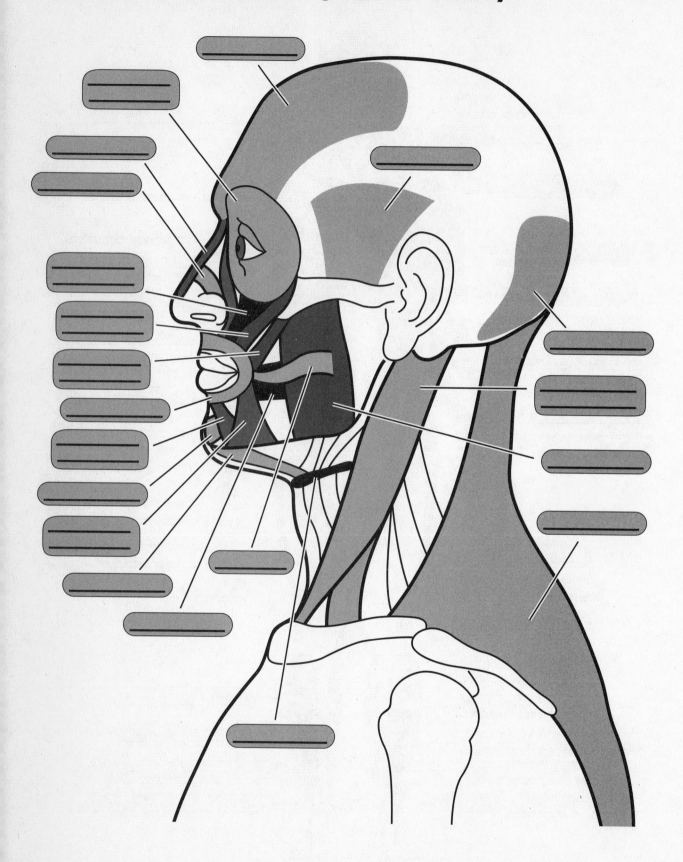

TORSO (anterior view)

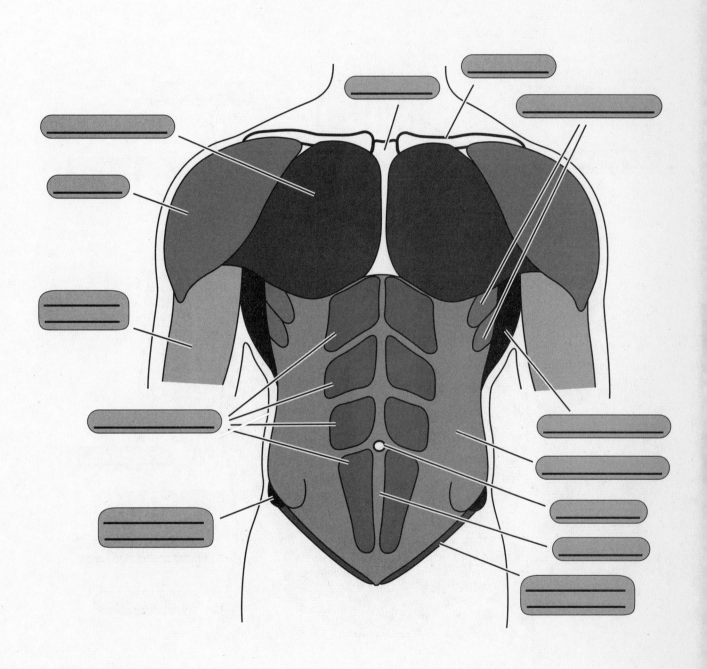

TORSO (posterior view)

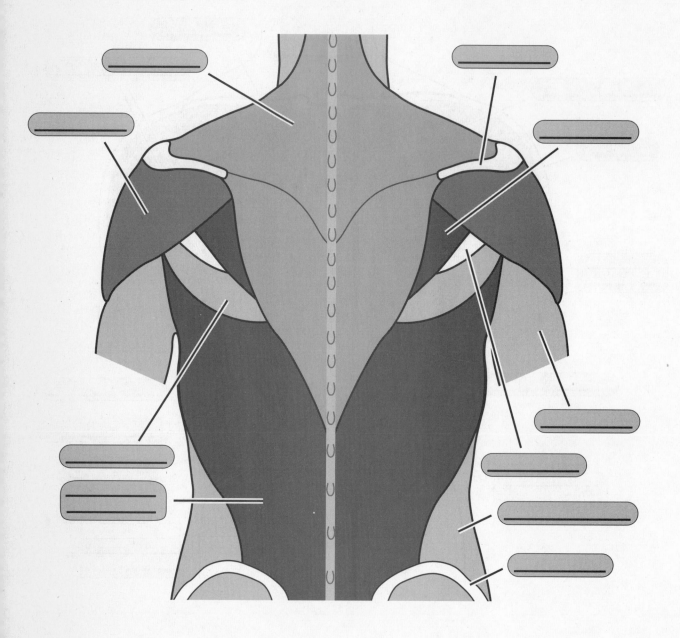

PERINEUM

Male
Perineum

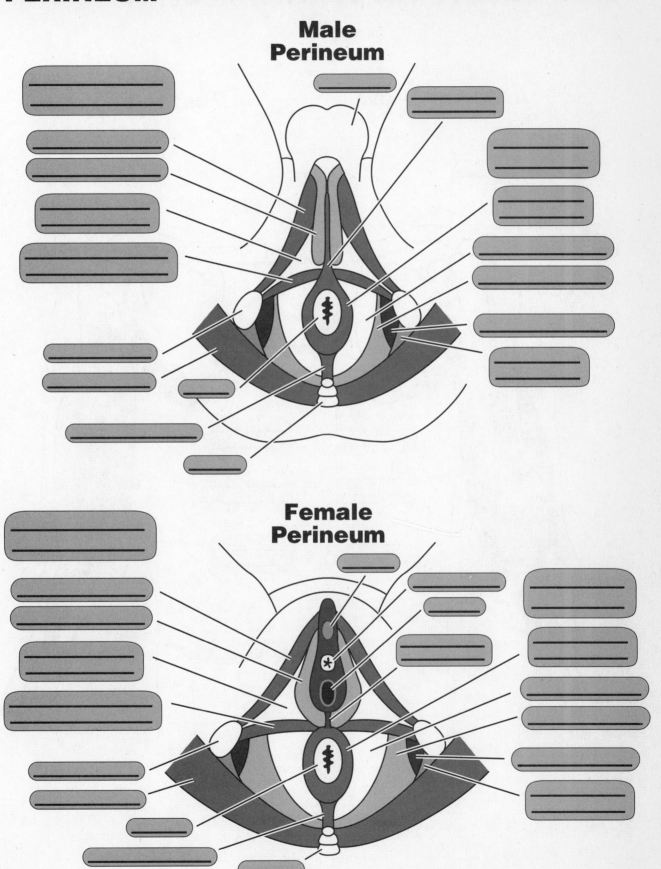

Female
Perineum

ARM (anterior and posterior views)

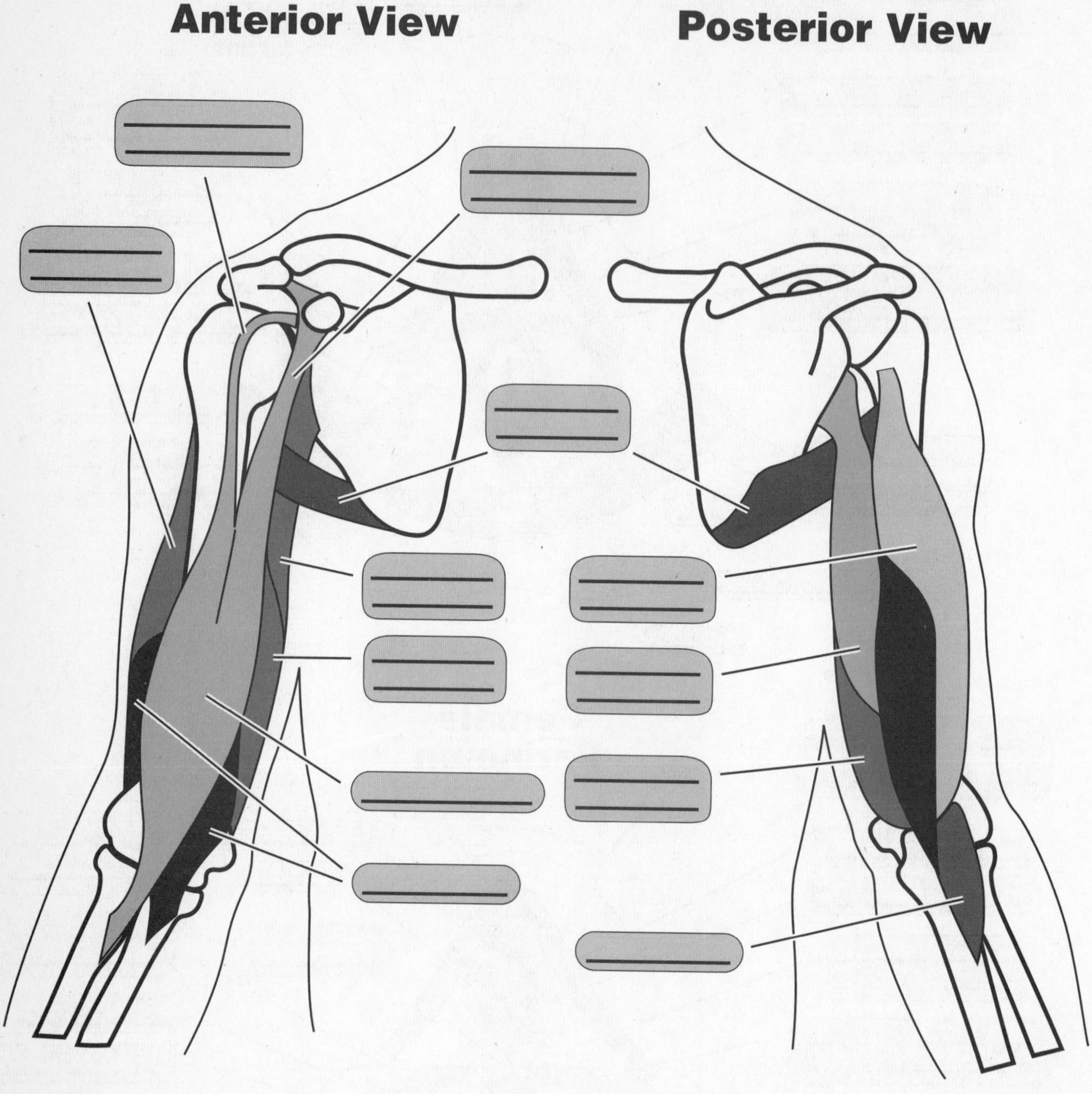

Anterior View

Posterior View

FOREARM (anterior view)

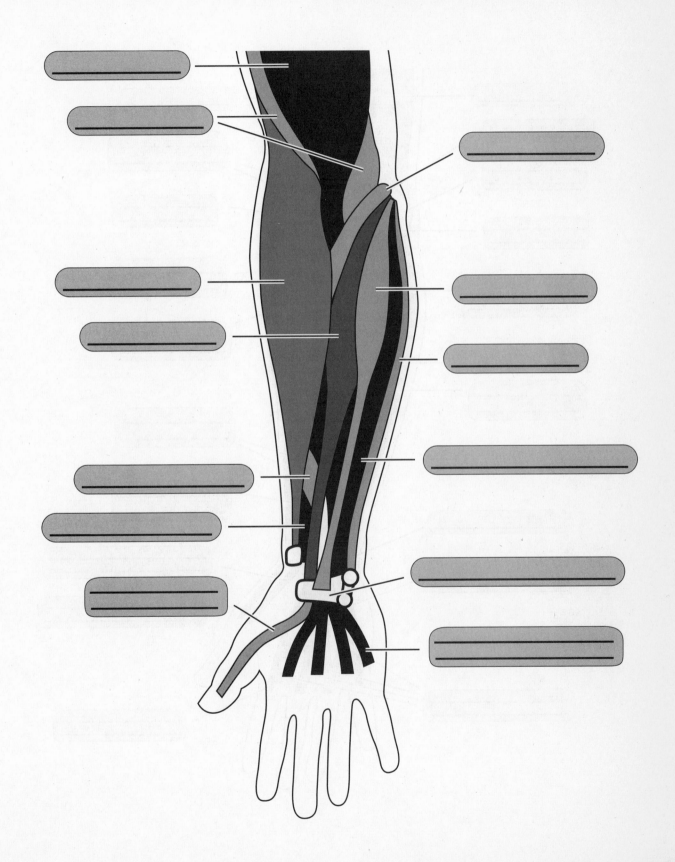

FOREARM (posterior view)

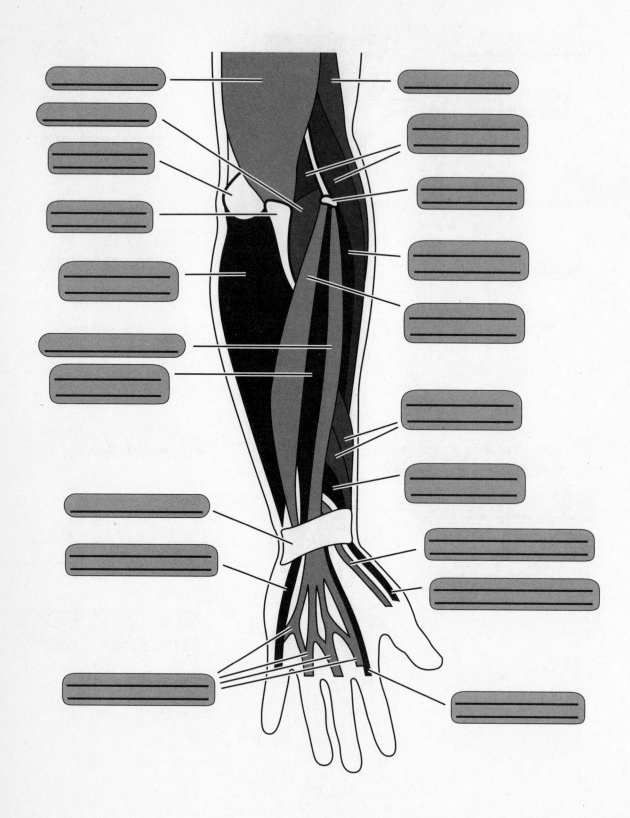

LOWER EXTREMITY (lateral and anterior views)

Lateral View

Anterior View

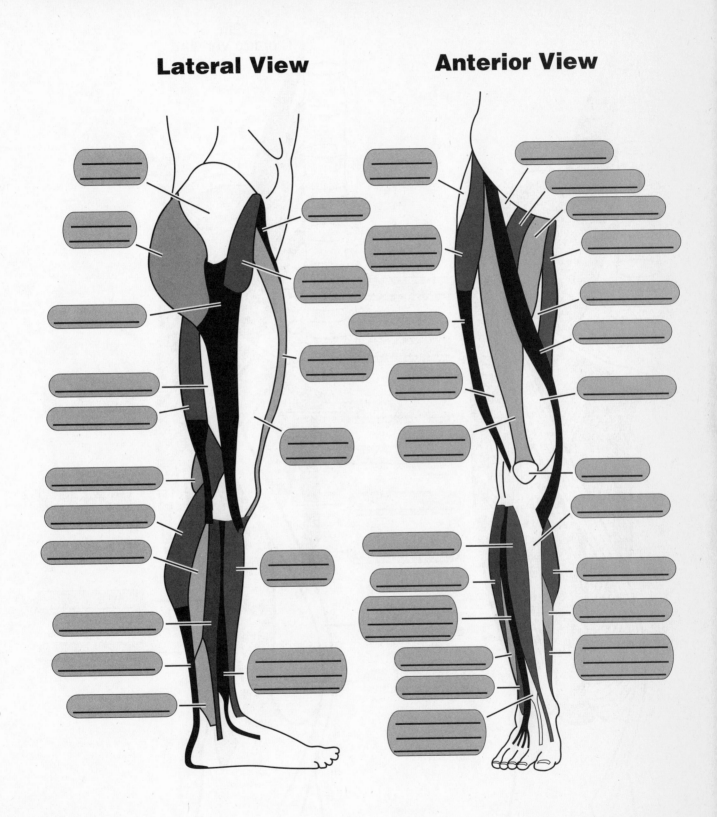

THIGH (anterior view)

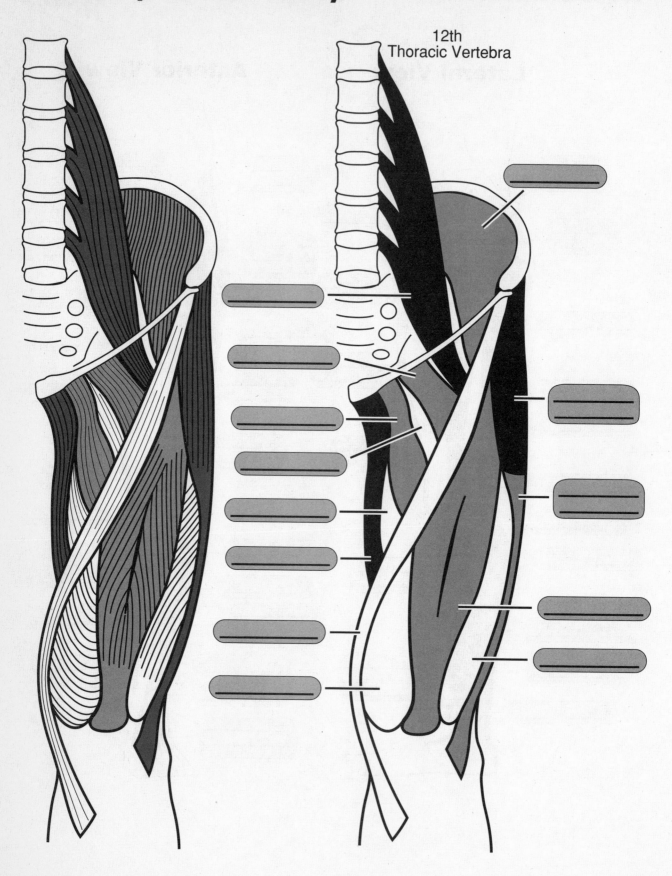

12th
Thoracic Vertebra

THIGH (posterior view)

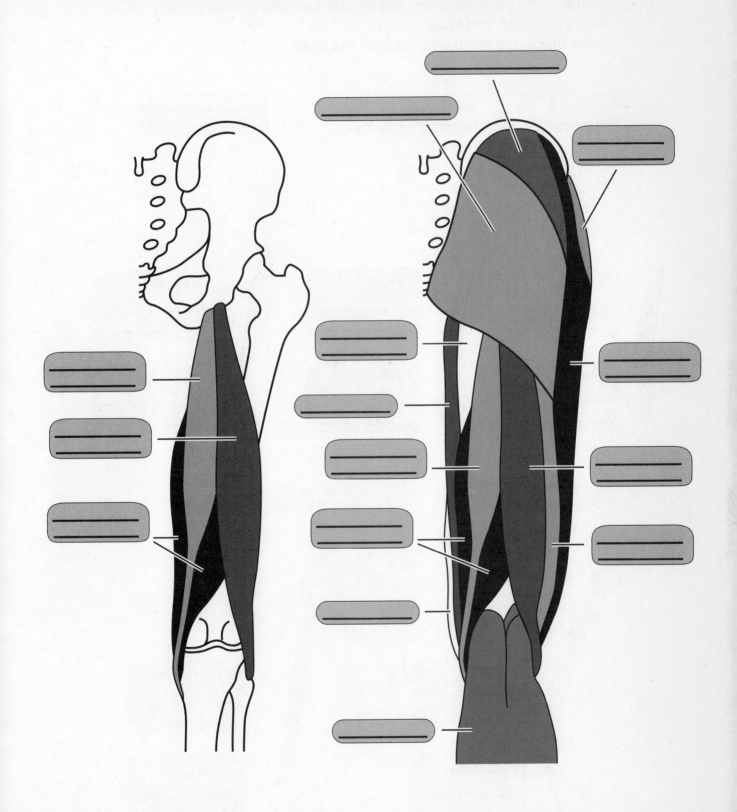

THIGH : anterior muscles

Vastus medialis, vastus intermedius, and vastus lateralis extend the leg.
Sartorius flexes the thigh and knee.
Rectus femoris flexes the thigh and extends the knee.

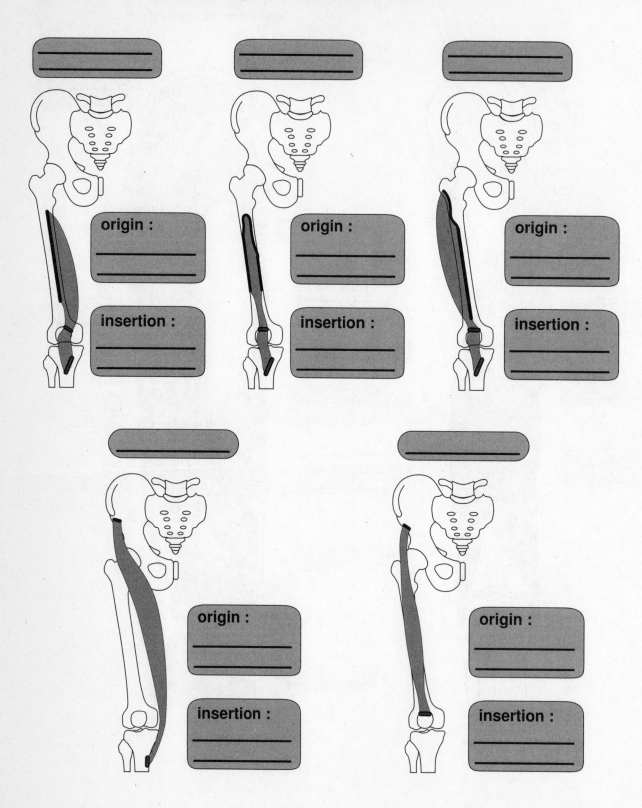

origin :

_____

_____

insertion :

_____

_____

origin :

_____

_____

insertion :

_____

_____

origin :

_____

_____

insertion :

_____

_____

origin :

_____

_____

insertion :

_____

_____

origin :

_____

_____

insertion :

_____

_____

THIGH : medial muscles

These muscles adduct and flex the thigh.
A pulled groin is the stretching or tearing of these muscles.

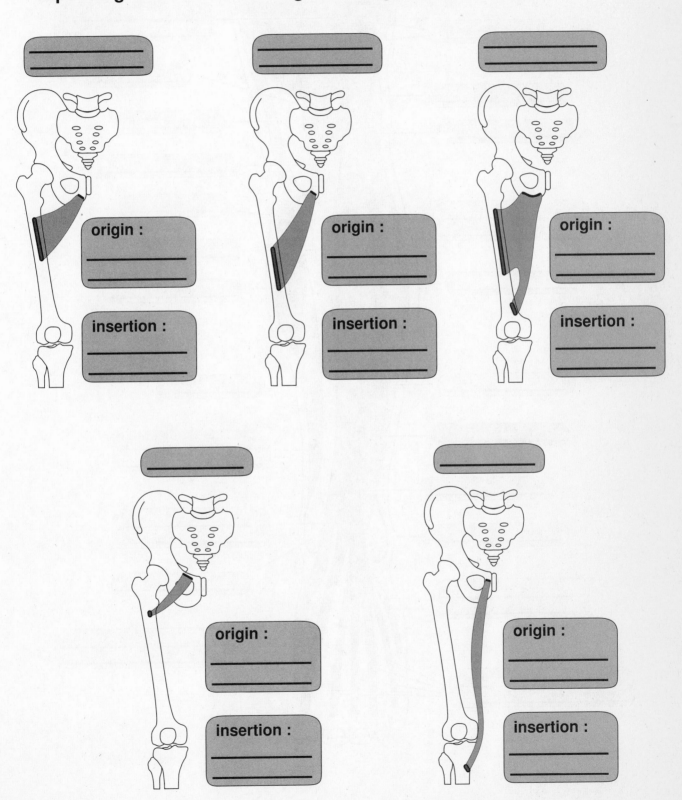

origin :

insertion :

origin :

insertion :

origin :

insertion :

origin :

insertion :

origin :

insertion :

LEG (anterior view)

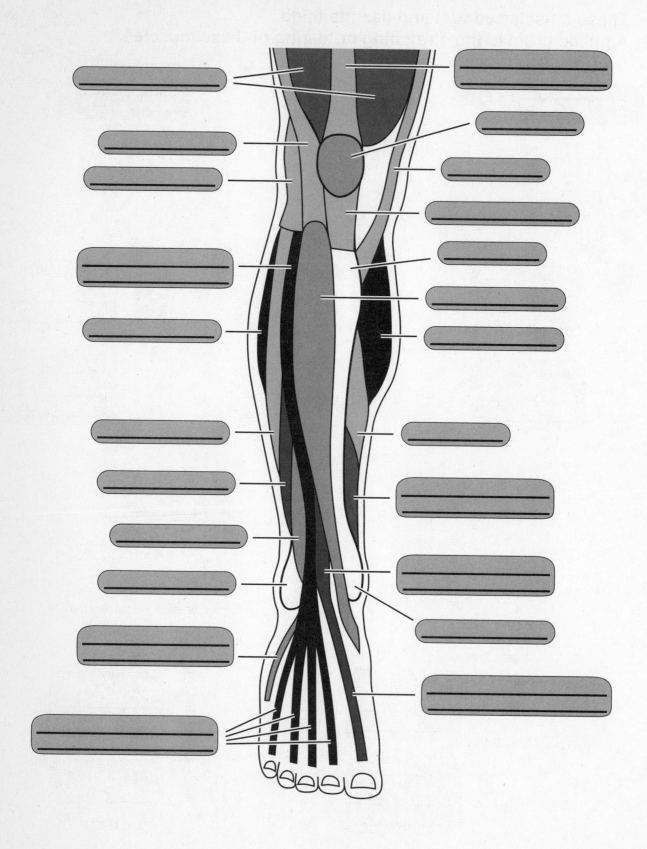

LEG (posterior view)

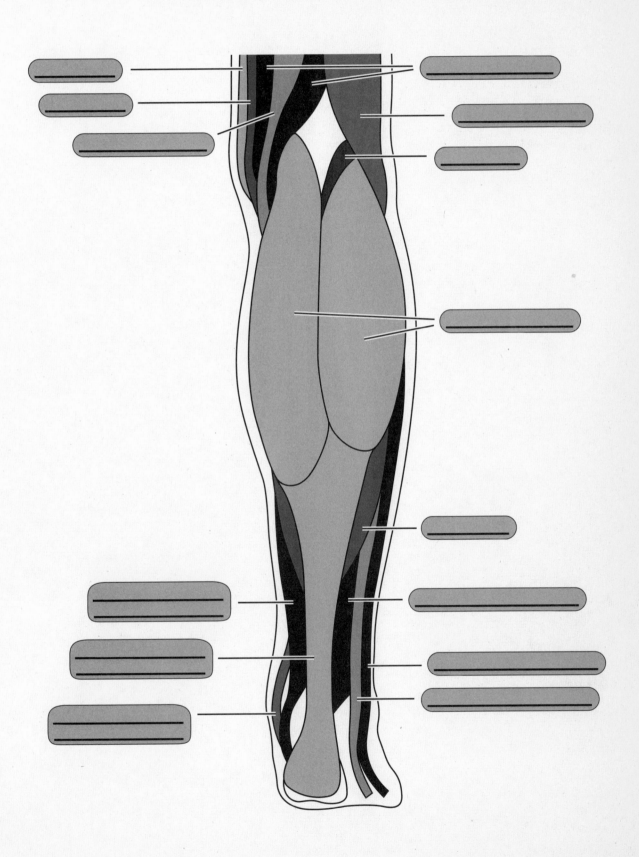

Part III : Terminology

Pronunciation Guide

Pronunciation Key

Accented Syllables

The strongest accented syllable is in capital letters : dī - ag - NŌ - sis
Secondary accents are indicated by a prime (′) : fiz′ - ē - OL - ō - jē

Vowels

Long vowels are indicated by a horizontal line (macron) above the letter :

ā as in make   ē as in be   ī as in ivy   ō as in pole

Short vowels are unmarked :

e as in pet   i as in pit   o as in pot   u as in bud

Other Phonetic Symbols

a as in about   oo as in do   yoo as in cute   oy as in noise

NOTE : Bones and skeletal muscles appear in separate lists at the end of this section.

abduction   ab - DUK - shun
abductor   ab - DUK - tor
acetabulum   as′ - e - TAB - yoo - lum
acetylcholine   as′ - e - til - KŌ - lēn
acetylcholinesterase   as′ - e - til - kō′ - lin - ES - ter - ās
Achilles   a - KIL - ēz
acromial   a - KRŌ - mē - al
acromion   a - KRŌ - mē - on
actin   AK - tin
adduction   ad - DUK - shun
adductor   ad - DUK - tor
adenosine triphosphate   a - DEN - ō - sen trī - FOS - fāt
aerobic   air - Ō - bik
agonist   AG - ō - nist
amphiarthroses   am′ - fē - ar - THRŌ - sēz
amphiarthrosis   am′ - fē - ar - THRŌ - sis
anaerobic   an - air - Ō - bik
annulus fibrosis   AN - yoo - lus fī - BRŌ - sus
anococcygeal raphe   ā′ - nō - kok - SIJ - ē - al RĀ - fē
antagonist   an - TAG - ō - nist
aponeurosis   ap′ - ō - noo - RŌ - sis
appendicular   ap′ - en - DIK - yoo - lar
appositional   a - pō - ZISH - a - nal
arthrodial   ar - THRŌ - dē - al
arthrosis   ar - THRŌ - sis
articular   ar - TIK - yoo - lar
articulation   ar - tik′ - yoo - LĀ - shun
autorhythmicity   aw′ - tō - rith - MIS - i - tē
axial   AK - sē - al
axon   AK - son

biceps   BĪ - seps

bipennate   bī - PEN - āt
boron   BOR - on
brevis   BREV - is
bursa   BUR - sa
bursae   BUR - sē

calcaneal   kal - KĀ - nē - al
calcification   kal - si - fi - KĀ - shun
calcitonin   kal - si - TŌ - nin
calcitriol   kal - si - TRĪ - ol
calmodulin   kal - MOD - yoo - lin
canaliculi   kan′ - a - LIK - yoo - lī
canaliculus   kan′ - a - LIK - yoo - lus
cancellous   KAN - sel - us
capitulum   ka - PIT - yoo - lum
cardiac   KAR - dē - ak
carotid   ka - ROT - id
cartilage   KAR - ti - lij
cartilaginous   kar′ - ti - LAJ - i - nus
cervical   SER - vi - kul
chondroblast   KON - drō - blast
chondrocyte   KON - drō - sīt
circumduction   ser′ - kum - DUK - shun
collagen   KOL - a - jen
concha   KONG - ka
conchae   KONG - kē
condyle   KON - dīl
condyloid   KON - di - loyd
contractility   kon′ - trak - TIL - i - tē
coracoid   KOR - a - koyd
coronal   kō - RŌ - nal
coronoid   KOR - ō - noyd

costal KOS-tal
creatine phosphate KRĒ-a-tin FOS-fāt
cribriform KRIB-ri-form
crista galli KRIS-ta GAL-lē

deltoid DEL-toyd
desmosome DEZ-mō-sōm
diaphysis dī-AF-i-sis
diarthroses dī'-ar-THRŌ-sēz
diarthrosis dī'-ar-THRŌ-sis
distal DIS-tal
dorsiflexion dor'-si-FLEK-shun

eccentric ek-SEN-trik
ectoderm EK-tō-derm
ellipsoidal e-lip-SOY-dal
endochondral en'-dō-KON-dral
endomysium en'-dō-MĪZ-ē-um
endosteum end-OS-tē-um
epicondyle ep'-i-KON-dīl
epimysium ep'-i-MĪZ-ē-um
epiphyseal ep'-i-FIZ-ē-al
epiphyses e-PIF-i-sēz
epiphysis e-PIF-i-sis
ethmoidal eth-MOY-dal
eversion ē-VER-zhun
extracapsular eks'-tra-KAP-syoo-lar

facet FAS-et
fascia FASH-ē-a
fascia lata FASH-ē-a LA-ta
fascicle FAS-i-kul
fasciculi fa-SIK-yoo-lī
fasciculus fa-SIK-yoo-lus
fibrous FĪ-brus
fibular FIB-yoo-lar
fissure FISH-ur
fixator fik-SĀ-tor
flexion FLEK-shun
flexor FLEK-sor
fontanel fon'-ta-NEL
foramen fō-RĀ-men
foramina fō-RAM-i-na
fossa FOS-a
fusiform FYOO-zi-form

gaster GAS-ter
ginglymus JIN-gli-mus
glenoid GLE-noyd
glucocorticoid gloo-kō-KOR-ti-koyd
gluteal GLOO-tē-al
glycogen GLĪ-kō-jen
glycolysis glī-KOL-i-sis
glycolytic glī'-kō-LIT-ik

Golgi GOL-jē
gomphosis gom-FŌ-sis

Haversian ha-VER-shun
hemopoiesis hē-mō-poy-Ē-sis
histology hiss-TOL-ō-jē
homeostasis hō-mē-ō-STĀ-sis
hydroxyapatite hī-drok'-sē-AP-a-tīt
hyperextension hī-per-ek-STEN-shun

iliac IL-ē-ak
iliotibial il'-ē-ō-TIB-ē-al
incisive in-SĪ-siv
inguinal IN-gwi-nal
insulin IN-su-lin
intercalated in-TER-ka-lāt-ed
intercondylar in'-ter-KON-di-lar
intercostal in'-ter-KOS-tal
interstitial in'-ter-STISH-al
intertrochanteric in'-ter-trō'-kan-TER-ik
intertubercular in'-ter-tyoo-BER-kyoo-lar
intervertebral in'-ter-VER-te-bral
intrafusal in'-tra-FYOO-zal
intramembranous in'-tra-MEM-bra-nus
inversion in-VER-zhun
ischial IS-kē-al
isometric ī'-sō-ME-trik
isotonic ī'-sō-TON-ik

jugular JUG-yoo-lar

kinase KĪ-nās
kinesthesia kin-is-THĒ-szē-a

lacrimal LAK-ri-mal
lactic LAK-tik
lacuna la-KOO-na
lacunae la-KOO-nē
lambdoid LAM-doyd
lamella la-MEL-a
lamellae la-MEL-ē
lamina LAM-i-na
laminae LAM-i-nē
latent LĀ-tent
levator le-VĀ-tor
ligament LIG-a-ment
linea alba LIN-ē-a AL-ba
linea aspera LIN-ē-a AS-pe-ra
longus LONG-gus
lumbosacral lum'-bō-SĀ-kral

magnum MAG-num
malleolus mal-LĒ-ō-lus
mandibular man-DIB-yoo-lar

manubrium   ma-NOO-brē-um
mastoid   MAS-toyd
matrix   MĀ-triks
maxillary   MAK-si-ler-ē
maximus   MAK-si-mus
meatus   mē-Ā-tus
medial   MĒ-dē-al
medullary   MED-yoo-lar'-ē
menisci   men-IS-ī
meniscus   me-NIS-kus
mesenchymal   MEZ-en-kīm-al
mesenchyme   MEZ-en-kīm
mesoderm   MEZ-ō-derm
mesodermal   mez'-ō-DER-mal
metaphysis   me-TAF-i-sis
minimus   MIN-i-mus
multipennate   mul'-ti-PEN-āt
myofibril   mī-ō-FĪ-bril
myoglobulin   mī-ō-GLŌ-bin
myology   mī-OL-ō-jē
myoneural   mī-ō-NOO-ral
myosin   MĪ-ō-sin

neuromuscular   noo-rō-MUS-kyoo-lar
neuron   NOO-ron
neurotransmitter   noo'-rō-TRANS-mit-er
nucleus pulposus   NOO-klē-us pul-PŌ-sus

oblique   ō-BLĒK
obturator   OB-too-rā'-ter
occipital   ok-SIP-i-tal
occipitomastoid   ok-sip'-i-tō-MAS-toyd
olecranon   ō-LEK-ra-non
olfactory   ōl-FAK-tō-rē
optic   OP-tik
oseous   OS-ē-us
ossification   os'-i-fi-KĀ-shun
osteoblast   OS-tē-ō-blast'
osteoclast   OS-tē-ō-clast'
osteocyte   OS-tē-ō-sīt'
osteogenic   os'-tē-ō-JEN-ik
osteon   OS-tē-on
osteoprogenitor   os'-tē-ō-prō-JEN-i-tor
ovale   ō-VAL-ē
oxidative   OK-si-dā-tiv

paranasal   par'-a-NĀ-zal
parathyroid   par'-a-THĪ-royd
patellar   pa-TEL-ar
patellofemoral   pa-tel'-ō-FEM-ō-ral
pectoral   PEK-tō-ral
pedicle   PED-i-kul
pelvic   PEL-vik
pennate   PEN-āt

perichondrium   per-i-KON-drē-um
perimysium   per-i-MĪZ-ē-um
perineum   per'-i-NĒ-um
periosteal   per'-ē-OS-tē-al
periosteum   per'-ē-OS-tē-um
phosphagen   FOS-fa-jin
phosphocreatine   fos'-fō-KRĒ-a-tin
phosphorus   FOS-fō-rus
plantar   PLAN-tar
prominence   PROM-i-nens
pronation   prō-NĀ-shun
pronator   prō-NĀ-tor
proprioception   prō-prē-ō-SEP-shun
protraction   prō-TRAK-shun
proximal   PROK-si-mal
pubic symphysis   PYOO-bik SIM-fi-sis

quadriceps   KWOD-ri-seps

rami   RĀ-mī
ramus   RĀ-mus
recruitment   rē-KROOT-ment
rectus   REK-tus
refractory   re-FRAK-to-rē
reticulum   re-TIK-yoo-lum
retinaculum   ret'-i-NAK-yoo-lum
retraction   rē-TRAK-shun
rhomboideus   rom-BOYD-ē-us
rotator   RŌ-tāt-or
rotundum   rō-TUN-dum

sacroiliac   sā'-krō-IL-ē-ak
sagittal   SAJ-i-tal
sarcolemma   sar'-kō-LEM-a
sarcomere   SAR-kō-mēr
sarcoplasmic   sar'-kō-PLAZ-mik
sciatic   sī-AT-ik
sellaris   sel-A-ris
sella turcica   SEL-a TUR-si-ka
serratus   ser-Ā-tis
sesamoid   SES-a-moyd
sphenofrontal   sfē'-nō-FRUN-tal
sphenoidal   sfē'-NOY-dal
sphenoparietal   sfē'-nō-pa-RĪ-e-tal
sphenosquamosal   sfē'-nō-skwā-MŌ-sal
spheroid   SFĒ-royd
sphincter   SFINGK-ter
spinosum   spī-NŌ-sum
spinous   SPĪ-nus
squamosal   skwā-MŌ-sal
squamous   SKWĀ-mus
striated   STRĪ-āt-ed
styloid   STĪ-loyd
subcutaneous   sub'-kyoo-TĀ-nē-us

sulci   SUL-sē
sulcus   SUL-kus
supination   soo'-pi-NĀ-shun
supinator   soo'-pi-NĀ-tor
supraorbital   soo'-pra-OR-bi-tal
suprapatellar   soo'-pra-pa-TEL-ar
sutural   SOO-chur-al
suture   SOO-cher
symphysis   SIM-fi-sis
synapse   SIN-aps
synaptic   sin-AP-tik
synarthroses   sin'-ar-THRŌ-sēz
synarthrosis   sin'-ar-THRŌ-sis
synchondrosis   sin'-kon-DRŌ-sis
syndesmosis   sin'-dez-MŌ-sis
synostosis   sin'-os-TŌ-sis
synovial   si-NŌ-vē-al

temporal   TEM-pō-ral
tendon   TEN-don
tensor   TEN-sor
testosterone   tes-TOS-te-rōn
tetanus   TET-a-nus
thermogenesis   ther'-mō-JEN-e-sis
thoracic   thō-RAS-ik
thorax   THŌ-raks
thyroid   THĪ-royd
thyroxine   thī-ROK-sin
tibial   TIB-ē-al
tibiofemoral   tib'-ē-ō-FEM-or-al
trabecula   tra-BEK-yoo-la
trabeculae   tra-BEK-yoo-lē
trapezius   tra-PĒ-zē-us
treppe   TREP-eh
triad   TRĪ-ad
triceps   TRĪ-seps
trochanter   trō-KAN-ter
trochlea   TRŌK-lē-Ā
trochlear   TRŌK-lē-ar
trochoid   TRŌ-koyd
tropomyosin   trō'-pō-MĪ-ō-sin
troponin   TRŌ-pō-nin
tubercle   TOO-ber-kul
tuberosity   too'-be-ROS-i-tē
turbinate   TUR-bi-nāt

ulnar   UL-nar'
unipennate   yoo-ni-PEN-āt

vertebral   VER-te-bral
vertebrochondral   ver'-te-brō-KON-dral
vertebrosternal   ver'-te-brō-STER-nal
visceral   VIS-er-al
Volkmann's   FŌLK-manz

wormian   WER-mē-an

xiphoid   ZĪ-foyd

zygomatic   zī-gō-MAT-ik

BONES

atlas   **AT - las**

auditory ossicles   **AW - di - tō - rē OS - si - kuls**

axis   **AK - sis**

calcaneus   **kal - KĀ - nē - us**

capitate   **KAP - i - tāt**

carpals   **KAR - pals**

carpus   **KAR - pus**

cervical vertebrae   **SER - vi - kul VER - te - brē**

clavicle   **KLAV - i - kul**

coccygeal vertebrae   **kok - SIJ - ē - al VER - te - brē**

coccyx   **KOK - six**

costa   **KOS - ta**

costae   **KOS - tē**

coxal   **KOK - sal**

cranial   **KRĀ - nē - al**

cranium   **KRĀ - nē - um**

cuboid   **KYOO - boyd**

cuneiform   **kyoo - NĒ - i - form**

ethmoid   **ETH - moyd**

facial   **FĀ - shal**

femur   **FĒ - mur**

fibula   **FIB - yoo - la**

frontal   **FRUN - tal**

hamate   **HAM - āt**

humerus   **HYOO - mer - us**

hyoid   **HĪ - oyd**

ilium   **IL - ē - um**

incus   **ING - kus**

innominate   **i - NOM - i - nāt**

inferior nasal concha   **NĀ - zal KONG - ka**

inferior nasal conchae   **NĀ - zal KONG - kē**

ischium   **IS - kē - um**

lacrimal   **LAK - ri - mal**

lumbar vertebrae   **LUM - bar VER - te - brē**

lunate   **LOO - nāt**

malleus   **MAL - ē - us**

mandible   **MAN - di - b'l**

maxilla   **mak - SIL - a**

maxillae   **mak - SIL - ē**

metacarpals   **met' - a - KAR - pals**

metacarpus   **met' - a - KAR - pus**

metatarsals   **met' - a - TAR - sals**

metatarsus   **met' - a - TAR - sus**

nasal   **NĀ - zal**

navicular   **na - VIK - yoo - lar**

occipital   **ok - SIP - i - tal**

os costae   **OS COS - tē**

os coxae   **OS KOK - sē**

palatine   **PAL - a - tīn**

parietal   **pa - RĪ - e - tal**

patella   **pa - TEL - a**

pectoral girdle   **PEK - tō - ral GIR - d'l**

pelvic girdle   **PEL - vik GIR - d'l**

phalanges   **fa - LAN - jēz**

phalanx   **FĀ - lanks**

pisiform   **PĪ - si - form**

pubis   **PYOO - bis**

radius   **RĀ - dē - us**

rib   **RIB**

sacral vertebrae   **SĀ - kral VER - te - brē**

sacrum   **SĀ - krum**

scaphoid   **SKAF - oyd**

scapula   **SKAP - yoo - la**

skull   **SKUL**

sphenoid   **SFĒ - noyd**

stapes   **STĀ - pēz**

sternum   **STER - num**

talus   **TĀ - lus**

tarsals   **TAHR - sals**

tarsus   **TAHR - sus**

temporal   **TEM - pō - ral**

thoracic vertebrae   **thō - RAS - ik VER - te - brē**

tibia   **TIB - ē - a**

trapezium   **tra - PĒ - zē - um**

trapezoid   **TRAP - ē - zoid**

triquetrum   **trī - KWĒ - trum**

turbinate   **TUR - bi - nāt**

ulna   **UL - na**

vertebra   **VER - te - bra**

vertebrae   **VER - te - brē**

vertebral column   **VER - te - bral**

vomer   **VŌ - mer**

zygomatic   **ZĪ - gō - mat' - ik**

SKELETAL MUSCLES

abductor pollicis longus　　ab-DUK-tor POL-li-kis LON-gus
adductor brevis　　ad-DUK-tor BREV-is
adductor longus　　ad-DUK-tor LONG-gus
adductor magnus　　ad-DUK-tor MAG-nus
anconeus　　an-KŌ-nē-us

biceps brachii　　BĪ-ceps BRĀ-kē-ī
biceps femoris　　BĪ-ceps FEM-or-is
brachialis　　brā′-kē-A-lis
brachioradialis　　brā′-kē-ō-rā′-dē-A-lis
buccinator　　BUK-si-nā′-tor
bulbocavernosus　　bul′-bō-ka′-ver-NŌ-sus

deep transverse perineus　　per-i-NĒ-us
deltoid　　DEL-toyd
depressor anguli oris　　de-PRE-ser ANG-gyoo-lī Or-is
depressor labii inferioris　　de-PRE-ser LA-bē-ī in-fer′-ē-OR-is
digastric　　dī′-GAS-trik

extensor carpi radialis brevis　　eks-TEN-sor KAR-pē rā′-dē-A-lis BREV-is
extensor carpi radialis longus　　eks-TEN-sor KAR-pē rā′-dē-A-lis LONG-gus
extensor carpi ulnaris　　eks-TEN-sor KAR-pē ui-NAR-is
extensor digiti minimi　　eks-TEN-sor DIJ-i-tē MIN-i-mē
extensor digitorum　　eks-TEN-sor dī′-ji-TOR-um
extensor digitorum longus　　eks-TEN-sor di′-ji-TOR-um LONG-gus
extensor hallucis longus　　eks-TEN-sor HAL-a-kis LONG-gus
extensor indicis　　eks-TEN-sor IN-di-kis
extensor pollicis brevis　　eks-TEN-sor POL-li-kis BREV-is
external anal sphincter　　eks-TER-nal Ā-nal SFINGK-ter
external oblique　　eks-TER-nal ō-BLĒK

flexor carpi radialis　　FLEK-sor KAR-pē rā′-dē-A-lis
flexor carpi ulnaris　　FLEK-sor KAR-pē ul-NAR-is
flexor digitorum longus　　FLEK-sor di′-ji-TOR-um LONG-gus
flexor digitorum superficialis　　FLEK-sor di′-ji-TOR-um soo′-per-fish′-ē-A-lis
flexor hallucis longus　　FLEK-sor HAL-a-kis LONG-gus
flexor pollicis longus　　FLEK-sor POL-li-kis LONG-gus
frontalis　　fron-TA-lis

gastrocnemius　　gas′-trok-NĒ-mē-us
gluteus maximus　　GLOO-tē-us MAK-si-mus
gluteus medius　　GLOO-tē-us MĒ-dē-us
gracilis　　gra-SIL-is

iliacus　　il′-ē-AK-us
iliococcygeus　　il′-ē-ō-kok-SIJ-ē-us
iliopsoas　　il′-ē-ō-SŌ-as
infraspinatus　　in′-fra-spi-NĀ-tus
ischiocavernosus　　is′-kē-ō-ka′-ver-NŌ-sus

latissimus dorsi　　la-TIS-i-mus DOR-sī
levator ani　　le-VĀ-tor Ā-nē
levator labii superioris　　le-VĀ-tor LA-bē-ī soo-per′-ē-OR-is

masseter   MA - se - ter
mentalis   men - TA - lis

nasalis   nā - ZA - lis

obturator internus   OB - too - rā′ - tor   in - TER - nus
occipitalis   ok - si′ - pi - TA - lis
orbicularis oculi   or - bi′ - kyoo - LAR - is   Ō - kyoo - lī
orbicularis oris   or - bi′ - kyoo - LAR - is   OR - is

palmaris longus   pal - MA - ris   LONG - gus
pectineus   pek - TIN - ē - us
pectoralis major   pek′ - tor - A - lis   MĀ - jor
peroneus brevis   per′ - ō - NĒ - us   BREV - is
peroneus longus   per′ - ō - NĒ - us   LONG - gus
peroneus tertius   per′ - ō - NĒ - us   TER - shus
plantaris   plan - TA - ris
procerus   pro - SE - rus
pronator quadratus   PRŌ - nā′ - ter   kwod - RĀ - tus
pronator teres   PRŌ - nā′ - ter   TE - rēz
psoas major   SŌ - as   MĀ - jor
pubococcygeus   pyoo′ - bō - kok - SIJ - ē - us

quadriceps femoris   KWOD - ri - seps   FEM - or - is

rectus abdominis   REK - tus   ab - DOM - in - us
rectus femoris   REK - tus   FEM - or - is
risorius   ri - ZOR - ē - us

sartorius   sar - TOR - ē - us
scalene   SKĀ - lēn
semimembranosus   sem′ - ē - mem′ - bra - NŌ - sus
semitendinosus   sem′ - ē - ten′ - di - NŌ - sus
serratus anterior   ser - Ā - tis   an - TĒR - ē - or
soleus   SŌ - lē - us
sternocleidomastoid   ster′ - nō - klī′ - dō - MAS - toyd
sternohyoid   ster - nō - HĪ - oyd
superficial perineus   per - i - NĒ - us

temporalis   tem′ - por - A - lis
tensor fasciae latae   TEN - sor   FA - shē - ē   LĀ - tē
teres major   TE - rēz   MĀ - jor
teres minor   TE - rēz   MĪ - nor
tibialis anterior   tib′ - ē - A - lis   an - TĒR - ē - or
tibialis posterior   tib′ - ē - A - lis   pos - TĒR - ē - or
trapezius   tra - PĒ - zē - us
triceps brachii   TRĪ - seps   BRĀ - kē - ī

vastus intermedius   VAS - tus   in′ - ter - MĒ - dē - us
vastus lateralis   VAS - tus   lat′ - er - A - lis
vastus medialis   VAS - tus   mē′ - dē - A - lis

zygomaticus major   zī - gō - MA - ti - kus   MĀ - jor
zygomaticus minor   zī - gō - MA - ti - kus   MĪ - nor

Glossary of Terms

A band   One of the transverse bands making up the repeating striated pattern of skeletal and cardiac muscle cells; located in the middle of a sarcomere.

Abduction   Movement away from the axis or midline of the body.

Accessory ligaments   Ligaments contained in many diarthroses (freely movable joints). There are two types : (1) extracapsular ligaments lie outside the articular capsule; (2) intracapsular ligaments are inside the articular capsule, but outside the synovial cavity.

Acetabulum   The large cup-shaped cavity on the lateral surface of the hipbone that receives the head of the femur.

Acetylcholine (ACh)   The neurotransmitter released by motor neurons at most neuromuscular junctions.

Acetylcholinesterase (AChE)   An enzyme that inactivates acetylcholine.

Achilles tendon   *See* Calcaneal tendon.

Actin   The contractile protein that makes up thin filaments in muscle fibers (cells).

Action potential   An electrochemical signal propagated over long distances by nerve and muscle cells. It is characterized by an all-or-none reversal of the membrane potential; the inside of the cell temporarily becomes positive relative to the outside.

Active tension   The tension generated by contractile elements (thin and thick filaments) in a sarcomere. Also called *internal tension*.

Adduction   Movement toward the axis or midline of the body.

Adenosine diphosphate (ADP)   The product formed when ATP is broken down to release energy; contains two phosphate groups.

Adenosine triphosphate (ATP)   The energy "currency" of all living cells; energy readily available for cellular activities. It is synthesized in a process called cellular respiration.

Adventitia   The outermost covering of a structure or organ.

Aerobic system   A system of energy (ATP) production; it is the complete oxidation of glucose by a process called cellular respiration. While slower than glycolysis, it yields much more energy, about 36 molecules of ATP from each glucose molecule. Slow oxidative and fast oxidative muscle fibers have a high capacity to generate ATP by the aerobic system. This process requires oxygen.

Afferent neuron   A neuron that carries impulses from sensory receptors into the CNS (brain or spinal cord). Also called a *sensory neuron*.

Agonist   *See* Prime mover.

All-or-none principle   An event occurs either maximally or not at all. In muscle physiology, muscle fibers (cells) of a motor unit contract to their fullest extent or not at all. In neuron physiology, if a stimulus is strong enough to initiate an action potential, a nerve impulse is transmitted along the entire neuron at a constant strength.

Alpha receptor   A type of receptor found on visceral effectors (cardiac muscle, smooth muscle, and glandular epithelium). Stimulation leads to excitation.

Alpha motor neuron   A large motor neuron which innervates skeletal muscle fibers.

Amphiarthrosis   A slightly movable joint (articulation). The articulating bony surfaces are separated by fibrous connective tissue or fibrocartilage.

Anaerobic respiration   Glycolysis. The sequence of reactions in cellular respiration that do not require the presence of oxygen. During this process one glucose molecule is converted into two pyruvic acid molecules and there is a net production of 2 molecules of ATP.

Anal triangle   The diamond-shaped perineum may be divided into anterior and posterior triangles by a line drawn between the ischial tuberosities. The posterior triangle is called the anal triangle and contains the anus.

Antagonist   A muscle that has an action opposite that of the prime mover (muscle that produces the desired motion).

Anterior   Nearer to the front; opposite of posterior. Also called *ventral*.

Annulus fibrosus   A ring of fibrous tissue and fibrocartilage that encircles the pulpy substance (nucleus pulposus) of an intervertebral disc.

Aponeurosis   A sheetlike tendon joining one muscle with another muscle or with a bone.

Appendicular skeleton   Bones of the pectoral girdles, the pelvic girdle, the upper and lower extremities.

Appositional growth   Growth due to surface deposition of material, as in the growth in diameter of cartilage and bone. Also called *exogenous growth*.

Arches (of feet)   Bones of the foot arranged in two arches to provide support and leverage, the longitudinal arch and the transverse arch; held together by a series of ligaments.

Arrector pili   Smooth muscles attached to hairs; contraction pulls the hairs into a more vertical position, resulting in "goose bumps."

Arthrodial joint   *See* Gliding joint.

Arthrology   The study or description of joints.

Arthrosis   A joint or articulation.

Articular capsule   Sleevelike structure around a synovial joint composed of a fibrous capsule and a synovial membrane.

Articular cartilage   Hyaline cartilage attached to articular bone surfaces.

Articular disc   Fibrocartilage pad between articular surfaces of bones of some synovial joints. Also called a *meniscus*.

Articulation   A joint. A point of contact between bones, between cartilage and bones, or between teeth and bones.

Atlas   The 1st cervical vertebra.

Auditory ossicle   One of the three small bones of the middle ear; either the malleus, incus, or stapes.

Autonomic nervous system (ANS)   Consists of neurons that carry nerve impulses between the central nervous system (brain and spinal cord) and cardiac muscle, smooth muscle, and glands. The two principal subdivisions are the sympathetic and parasympathetic nervous systems.

Axial skeleton   The skull, hyoid bone, vertebral column, and rib cage.

Axis   The 2nd cervical vertebra.

Axon   The usually single, long process of a neuron that carries a nerve impulse away from the neuron cell body.

Axon terminal   Terminal branch of an axon.

Ball-and-socket   A synovial joint in which the rounded surface of one bone moves within a cup-shaped depression or fossa of another bone. Examples include the shoulder joint and hip joint. Also called a *spheroid joint*.

Base   The broadest part of a pyramidal bone. For example, each metacarpal bone, metatarsal bone, and phalanx has a proximal base.

Belly   The prominent, fleshy part of a skeletal muscle. Also called the *gaster*.

Beta receptor   A type of receptor found on visceral effectors (cardiac muscle, smooth muscle, and glandular epithelium). Stimulation leads to inhibition.

Body   The entire material structure of an animal, as in the human body; the main part of anything. Examples include: the cubelike central portion of the sphenoid bone; the curved, horizontal portion of the mandible; the horizontal portion of the hyoid bone; the thick, disc-shaped anterior portion of a vertebra; the middle, largest portion of the sternum; the main part of the femur, humerus, and ribs. The body of a vertebra is also called the *centrum*. The body of the femur, humerus, and ribs is also called the *shaft*.

Bone tissue   A type of connective tissue. The two main kinds of bone tissues are compact bone and spongy bone. Also called *osseous tissue*.

Bone types   On the basis of shape, bones are classified as long, short, flat, or irregular.

Bone surface markings   Terms that describe bone markings include fissure, foramen, meatus, fossa, process, condyle, head, facet, tuberosity, crest, and spine. Each marking is structured for a specific function—joint formation, muscle attachment, or passage of nerves and blood vessels.

Bursa (plural : bursae)   A sac or pouch of synovial fluid located at friction points, especially about joints.

Buttocks   The two fleshy masses on the posterior aspect of the lower trunk, formed by the gluteal muscles.

Calcaneal tendon   The tendon of the soleus, gastrocnemius, and plantaris muscles at the back of the heel. Also called the *Achilles tendon*.

Calcification   Deposition of mineral salts, primarily hydroxyapatite, in a framework formed by the collagen fibers of the matrix; crystallization occurs and the tissue hardens. Also called *mineralization*.

Calcitonin   A hormone produced by the thyroid gland that lowers the calcium and phosphate levels of the blood by inhibiting bone breakdown and accelerating calcium absorption by bones.

Callus   A growth of new bone tissue in and around a fractured area; ultimately replaced by mature bone.

Calmodulin   A regulator protein present in muscle cells that binds calcium.

Canal   A narrow tube, channel, or passageway in a bone. An example is the external auditory canal. Also called *meatus*.

Canaliculi (singular : canaliculus)   Small channels or canals in compact bone. They house the filopodial processes of osteocytes and facilitate the transport of materials between these mature bone cells.

Cancellous bone   Having a reticular or latticework structure, as in spongy bone.

Cardiac muscle tissue   Heart muscle cells; characterized by branching, striated fibers attached by intercalated discs.

Carpus   A collective term for the 8 bones of the wrist.

Cartilage   A type of connective tissue. Consists of cells called chondrocytes, which are located in spaces called lacunae; the matrix consists of a dense network of collagenous and elastic fibers and a ground substance of chondroitin sulfate.

Cartilaginous joint   A joint without a synovial cavity where the articulating bones are held tightly together by cartilage, allowing little or no movement.

Center of ossification   An area in the cartilage model of a future bone where the cartilage cells enlarge and then secrete enzymes that trigger the calcification of the matrix. The cartilage cells die and the area is invaded by osteoblasts that lay down bone.

Central canal   Found in compact bone tissue. A circular channel running longitudinally in the center of an osteon (Haversian system), containing blood and lymphatic vessels and nerves. Also called an *Haversian canal*.

Cervical vertebrae   The 7 vertebrae in the neck.

Chondroblast   Cell that forms cartilage.

Chondrocyte   Cell of mature cartilage.

Ciliary muscle   Smooth (involuntary) muscle in the eye that controls the shape of the lens.

Circumduction   A movement at a synovial joint; the distal end of a bone moves in a circle while the proximal end remains relatively stable.

Coccyx   The 4 fused vertebrae at the end of the vertebral column. Also called the *tailbone*.

Collagen   A type of protein fiber found in most types of connective tissues, especially bone, cartilage, tendons, and ligaments; gives strength to connective tissue.

Compact bone   Bone tissue with no apparent spaces in which the layers of lamellae are fitted tightly together. It is found immediately deep to the periosteum and external to spongy bone. Also called *dense bone*.

Complete tetanus   Muscle stimulation at a rate of 80 to 100 stimuli per second results in sustained contraction that lacks even partial relaxation between stimuli. Also called *fused tetanus*.

Concentric contraction   The action of the prime mover. For example, during flexion of the arm the biceps brachii is the prime mover; its action is called concentric contraction.

Concha (plural : conchae)   A scroll-like bone found in the skull.

Conducting system   A network of cardiac muscle fibers specialized to conduct electrical impulses to different areas of the heart.

Conductivity   The ability to carry the effect of a stimulus from one part of a cell to another; highly developed in nerve and muscle fibers (cells).

Condyle   A large, rounded articular prominence. An example is the medial condyle of the femur.

Condyloid joint   *See* Ellipsoidal joint.

Connective tissue   The most abundant of the four basic tissue types in the body. It binds together, supports, and strengthens other tissues; protects and insulates internal organs; and compartmentalizes structures such as skeletal muscles.

Constrictor muscle   Circular, smooth muscle (involuntary) in the iris of the eye. Contraction decreases the diameter of the pupil.

Contractility   The ability of cells or parts of cells to actively generate force. Muscle fibers exhibit a high degree of contractility.

Contraction   Activation of the tension-generating process that results in a shortening of a muscle fiber or whole muscle.

Contraction period The time interval between a muscle action potential and the development of peak tension or shortening by a muscle.

Costal Pertaining to a rib.

Costal cartilage Hyaline cartilage that attaches a rib to the sternum.

Coxal bones Hipbones. The pelvic girdle consists of the two coxal bones. Each coxal bone consists of three fused bones: ilium, ischium, and pubis. Also called the *os coxae* and the *innominate bones*.

Cramp A spasmodic muscular contraction; usually painful. Especially contractions characterized by continuous tension (tonic spasms).

Cranial bones The 8 cranial bones include the frontal, occipital, sphenoid, ethmoid, parietal (2), and temporal (2).

Creatine phosphate High-energy molecule in skeletal muscle fibers that is used to generate ATP rapidly. On decomposition, creatine phosphate breaks down into creatine, phosphate, and energy—the energy is used to generate ATP from ADP. Also called *phosphocreatine*.

Cross-bridge A projection extending from a thick filament in muscle. A portion of the myosin molecule capable of exerting force on the thin filament and causing it to slide past the thick one.

Deep Away from the surface of the body.

Deep fascia A sheet of connective tissue wrapped around a muscle to hold it in place.

Deep inguinal ring A slitlike opening in the aponeurosis of the transversus abdominis muscle; it is the origin of the inguinal canal.

Demineralization Loss of calcium and phosphorus from bones.

Dense bodies Structures in smooth muscle fibers that are attached to intermediate filaments; they are similar to the Z discs in striated muscle fibers.

Depression Movement in which a part of the body moves downward.

Desmosome Type of cell junction whose function is to hold cells together.

Diaphragm The dome-shaped sheet of skeletal muscle which separates the abdominal and thoracic cavities; the principal muscle of respiration.

Dilator muscle Radial, smooth muscle (involuntary) in the iris of the eye. Contraction increases the diameter of the pupil.

Diaphysis The shaft of a long bone.

Diarthrosis A freely movable joint, as in a hinge joint.

Distal Farther from the point of origin; toward the hand of the upper extremity or toward the foot of the lower extremity.

Dorsal Near to the back; opposite of ventral. Also called *posterior*.

Dorsiflexion Bending the foot or hand in the direction of the dorsum (upper surface).

Effector A muscle or a gland, that responds to instructions received from a control center. One of the three basic components of a feedback system (reflex arc).

Efferent neuron A neuron that carries impulses away from the brain or spinal cord toward an effector (muscle or gland). Also called a *motor neuron*.

Effort (E) In a lever system, effort is the force exerted to achieve an action. (*Resistance* is the force that opposes movement.)

Elasticity The ability of tissue to return to its original shape after contraction or extension.

Elevation Movement in which a part of the body moves upward.

Ellipsoidal joint A synovial joint structured so that an oval-shaped condyle of one bone fits into an elliptical cavity of another bone, permitting side-to-side and back-and-forth movements. An example is the joint at the wrist between the radius and carpals. Also called a *condyloid joint*.

Endochondral ossification The replacement of cartilage by bone. Also called *intracartilaginous ossification*.

Endomysium Extensions of deep fascia (dense, irregular connective tissue) that surround indvidual muscle fibers.

Endosteum The membrane that lines the medullary cavity of bones. It consists of osteoprogenitor cells and scattered osteoclasts.

Epicondyle A prominence above a condyle. An example is the medial epicondyle of the femur.

Epimysium Extensions of deep fascia (dense, irregular connective tissue) that surround a muscle.

Epiphyseal line The remnant of the epiphyseal plate in a long bone.

Epiphyseal plate The cartilaginous plate between the epiphysis and diaphysis. It is responsible for the lengthwise growth of long bones.

Epiphysis The end of a long bone, usually larger in diameter than the shaft (diaphysis).

Eversion The movement of the sole of the foot outward at the ankle joint.

Excitability The ability of muscle tissue to receive and respond to stimuli. The ability of nerve cells to respond to stimuli and convert them into nerve impulses.

Extensibility The ability of muscle tissue to be stretched when pulled.

Extension An increase in the angle between two bones. Restoring a body part to its anatomical position after flexion.

Extensor Any muscle that causes extension of a body part.

Extensor retinaculum A ligament in the wrist. It is located over the dorsal surface of the carpal bones and is attached to the styloid of the ulna, the radius, and two carpals. Through it pass the extensor tendons of the digits and wrist. Also called the *dorsal carpal ligament*.

Extrinsic Of external origin.

Extrinsic muscles Muscles that originate outside of the structure on which they act. Examples include : extrinsic eye muscles (voluntary muscles attached to the outside of the eyeball) and extrinsic muscles of the tongue (originate outside the tongue and insert into it).

External Located on or near the surface.

Facet A smooth, flat surface on a bone that forms part of a joint. An example is an articular facet on a vertebra.

Facial Pertaining to the face.

Facial bones The 14 facial bones include the mandible, vomer, maxillae (2), zygomatic (2), nasal (2), palatine (2), lacrimal (2), and inferior nasal conchae (2).

False ribs The 8th through the 12th pair of ribs. Their costal cartilages either attach indirectly to the sternum or do not attach to the sternum at all. The 11th and 12th pairs are sometimes called *floating ribs*, because they have no cartilaginous attachments to the sternum.

Fascia   A fibrous membrane covering, supporting, and separating muscles. Located outside the epimysium.

Fascia lata   A deep fascia of the thigh that encircles the entire thigh. Together with the tendons of the gluteus maximus and tensor fasciae latae muscles it forms a structure called the iliotibial tract.

Fascicle   A bundle of muscle fibers or nerve fibers. Also called *fasciculus* (plural : fasciculi).

Fascicular arrangements   The arrangement of skeletal muscle fasciculi with respect to their tendons take several characteristic patterns. They include parallel, convergent, pennate, and circular.

Fast glycolytic fibers (type II B fibers)   Muscle fibers that have a low myoglobin content, relatively few mitochondria, and relatively few capillaries. They contain large amounts of glycogen and are geared to generate ATP by anaerobic processes (glycolysis). They are the largest diameter fibers and are white. Also called *fast-twitch B* or *fatigable fibers*.

Fast oxidative fibers (type II A fibers)   Muscle fibers that contain large amounts of myoglobin, many mitochondria, and many capillaries. They are red, have a high capacity for generating ATP by oxidative processes, and are resistant to fatigue. Also called *fast-twitch A* or *fatigue resistant fibers*.

Fatigue   *See* Muscle fatigue.

Femoral   Pertains to the thigh; region between the groin and the knee.

Fibroblast   A large, flat cell that secretes the matrix of loose, connective tissue. The matrix includes the fibers and the ground substance.

Fibrocyte   A mature fibroblast that no longer produces fibers or ground substance in connective tissue.

Fibrous joint   A joint that allows little or no movement, such as a suture (joint in the skull).

Filopodial processes   Cytoplasmic extensions of osteocytes. They pass through minute channels called canaliculi and form gap junctions with the filopodial processes of osteocytes in adjacent lacunae.

Fissure   A deep depression or groove that may be normal or abnormal.

Fixator   A muscle that stabilizes the origin of the prime mover so that the prime mover can act more efficiently.

Flexion   A folding movement in which there is a decrease in the angle between two bones.

Flexor muscle   A muscle that causes flexion.

Flexor retinaculum   A ligament in the wrist. It is located over the palmar surface of the carpal bones. Through it pass the long flexor tendons of the digits and wrist and the median nerve. Also called the *transverse carpal ligament*.

Floating rib   A rib that does not attach to the sternum. The last two pairs of ribs (11th and 12th).

Fontanel   A membrane-covered spot where bone formation is not yet complete. Especially between the cranial bones of an infant's skull.

Foot   The terminal part of the lower extremity; includes the ankle (tarsus), arch or sole (metatarsus), and toes (phalanges).

Foramen   (plural: foramina)   A passage or opening; a communication between two cavities of an organ or a hole in a bone for passage of blood vessels or nerves.

Forearm   The part of the upper extremity between the elbow and the wrist.

Fossa   A furrow or shallow depression.

Fulcrum   In a lever system, the fulcrum is a fixed point, like the central point of attachment of a seesaw.

Gap junction   A type of cell junction that allows ions and small molecules to flow between adjacent cells. Small channels filled with cytoplasm connecting adjacent cells.

Ginglymus joint   *See* Hinge joint.

Gliding joint   A synovial joint having articulating surfaces that are usually flat, permitting side-to-side and back-and-forth movements. Between carpal bones, tarsal bones, and the scapula and clavicle. Also called an *arthrodial joint*.

Glycogen   A highly branched polymer of glucose containing thousands of subunits; stored in liver and muscle cells as an energy reserve.

Glycogen-lactic acid system   When there is not enough oxygen present pyruvic acid produced by glycolysis is converted into lactic acid. Heart muscle fibers, kidney cells, and liver cells can use lactic acid to produce ATP; liver cells convert some of the lactic acid back to glucose. Some lactic acid accumulates in blood and muscle tissue.

Glycolysis   The breakdown of a molecule of glucose into two molecules of pyruvic acid, yielding 2 ATP. Also called *anaerobic respiration*.

Golgi tendon organ   See *Tendon organ*.

Gomphosis   A fibrous joint in which a cone-shaped peg fits into a socket; articulating bones separated by periodontal ligament. An example is the roots of teeth in sockets.

Groin   The depression between the thigh and the trunk; the inguinal region.

Ground substance   The material that surrounds cells and fibers in connective tissue. It is secreted by fibroblasts and can be watery or gel-like. Ground substance together with the fibers constitute the matrix of connective tissues.

Growth   An increase in size. Due to an increase in the number of cells, the size of existing cells, or the amount of matrix.

Ham strings   Tendons of the semimembranosus, semitendinosus, gracilis, sartorius, and biceps femoris muscles.

Hand   The terminal portion of an upper extremity; includes the wrist (carpus), palm (metacarpus), and fingers (phalanges).

Hard palate   The anterior portion of the roof of the mouth, formed by the maxillae and palatine bones and lined by mucous membrane.

Haversian canal   *See* Central canal.

Haversian system   *See* Osteon.

Head   The superior part of a human, cephalic to the neck. The proximal end of a long bone. The part of a muscle attached to the less movable part of the skeleton; the muscle origin.

Hematopoiesis   *See* Hemopoiesis.

Hemopoiesis   Blood cell production occurring in the red marrow of bones. Also called *hematopoiesis*.

Hinge joint   A synovial joint in which a convex surface of one bone fits into a concave surface of another bone. Examples include the elbow, knee, ankle, and interphalangeal joints. Also called a *ginglymus joint*.

Hip bone   *See* Coxal bone.

Human growth hormone (hGH)   Hormone secreted by the anterior pituitary gland. It stimulates the growth of body tissues, especially skeletal and muscular tissues. Also called *somatotropin* and *somatotropic hormone (STH)*.

Hyaluronic acid   A viscous, amorphous extracellular material that binds cells together, lubricates joints, and maintains the shape of the eyeballs.

Hydroxyapatite (tricalcium phosphate)   Principal mineral salt in the matrix of bone.

Hyoid bone   A U-shaped bone in the neck. It is unique because it does not articulate with any other bone. It supports the tongue and provides attachment for some of its muscles and for muscles of the neck and pharynx.

Hormones   Chemicals secreted by endocrine glands that circulate in the bloodstream and alter the activities of their target cells. Hormones important for muscle and bone functions include parathyroid hormone (PTH), human growth hormone (hGH), calcitonin, thyroid hormones, and sex hormones. PTH <u>increases</u> blood calcium; calcitonin <u>decreases</u> blood calcium.

Hyperextension   Continuation of extension beyond the anatomical position.

Illiotibial tract   A structure located on the lateral surface of the thigh; it is composed of the fascia lata and the tendons of the gluteus maximus and tensor fasciae latae muscles. The tract inserts into the lateral condyle of the tibia.

Incomplete tetanus   Partial relaxation of a muscle between stimuli during sustained contraction. Results when muscle is stimulated at a rate of 20 to 30 stimuli per second. Also called *unfused tetanus*.

Inferior   Away from the head or toward the lower part of a structure. Also called *caudad*.

Inguinal   Pertaining to the groin (the depression between the thigh and the trunk).

Inguinal canal   An oblique passageway in the anterior abdominal wall just superior and parallel to the medial half of the inguinal ligament. In the male it is the location of the spermatic cord and the ilioinguinal nerve; in the female it is the location of the round ligament and the ilioinguinal nerve.

Innominate bones   *See* Coxal bones.

Insertion   The attachment of a muscle tendon to a movable bone; the end of a muscle that is opposite its origin.

Intercalated disc   An irregular transverse thickening of plasma membrane between attached heart muscle cells. Contains desmosomes and gap junctions. Desmosomes hold cardiac muscle cells together; gap junctions aid in conduction of muscle action potentials.

Intercostal nerve   A nerve supplying a muscle located between the ribs.

Internal   Away from the surface of the body.

Intervertebral disc   A pad of fibrocartilage located between the bodies of two vertebrae.

Intrafusal fibers   Muscle fibers located in muscle spindles (receptors in skeletal muscle that are sensitive to changes in length or tension). Three to ten specialized muscle fibers partially enclosed in a spindle-shaped connective tissue capsule that is filled with lymph.

Intramembranous ossification   The method of bone formation in which the bone is formed directly in membra-nous tissue.

Intrinsic   Situated entirely within a structure.

Intrinsic muscles   Muscles that are inside the structures on which they act. Examples include : the intrinsic eye muscles (the ciliary muscle controls the shape of the lens; the dilator and constrictor muscles control the diameter of the pupil) and the intrinsic muscles of the tongue (alter the shape and size of the tongue for speech and swallowing).

Inversion   The movement of the sole inward at the ankle joint.

Involuntary muscle   Muscles that are not under conscious, voluntary control. Smooth and cardiac muscle.

Isometric contraction   A contraction in which the muscle maintains the same length (metric); the tension increases.

Isotonic contraction   A contraction in which the muscle maintains the same tension (tonic); the length shortens.

Joint   A point of contact between bones, cartilage and bones, or teeth and bones. Also called *articulation* and *arthrosis*.

Joint kinesthetic receptor   A proprioceptive receptor located in a joint; stimulated by joint movement.

Kinesiology   The study of the movement of body parts.

Kinesthesia   The awareness of the directions of movement.

Knee   Tibiofemoral joint. A freely movable joint (diarthrosis) that consists of three joints : (1) intermediate patellofemoral joint (gliding joint); (2) lateral tibiofemoral joint (hinge joint); and (3) medial tibiofemoral joint (hinge joint).

Lacuna   (plural : lacunae)   A small, hollow space found in compact bone tissue; location of mature bone cells called osteocytes.

Lambdoid suture   The line of union in the skull between the parietal bones and the occipital bone; sometimes contains sutural bones (Wormian bones).

Lamellae (singular : lamella)   Concentric rings of hard, calcified matrix found in compact bone.

Lateral   Farther from the midline of the body or a structure.

Leg   The part of the lower extremity between the knee and the ankle.

Lever   A rigid rod that moves about on some fixed point called a fulcrum. Its purpose is to reduce the effort required to move a weight (resistance).

Ligament   Dense, regular, connective tissue that attaches bone to bone.

Long bone   One of four principal types of bone. Long bones have greater length than width and consist of a shaft and a variable number of ends.

Lower extremity   The appendage attached at the pelvic (hip) girdle, consisting of the thigh, knee, leg, ankle, foot, and toes.

Lumbar   Region of the back and side between the ribs and pelvis; the loin.

Lumbar vertebrae   Five vertebrae in the lower back; between the thoracic vertebrae and the sacrum.

Lumen   The space within a hollow tube or organ; as in the lumen of the intestine.

Marrow   Soft, spongelike material in the cavities of bone. Red marrow produces blood cells; yellow marrow stores fat.

Matrix   The ground substance and fibers that surround the cells in connective tissues. The matrix of bone contains abundant mineral salts (mostly hydroxyapatite) and collagen fibers.

Meatus   *See* Canal.

Medullary cavity   The space within the shaft (diaphysis) of a bone that contains yellow marrow. Also called the *marrow cavity*.

Membrane   A thin, flexible sheet of tissue. An epithelial membrane is composed of an epithelial layer and an underlying connective tissue layer; a synovial membrane is composed of areolar connective tissue only.

Meniscus   *See* Articular disc.

Mesenchyme   An embryonic connective tissue from which all other connective tissues arise.

Mesoderm   The middle of the three primary germ layers that gives rise to connective tissues and muscles.

Metacarpal   Pertaining to the metacarpus (the five bones of the hand between the carpals and phalanges); the palm.

Metaphysis   Growing portion of a bone.

Metatarsal   Pertaining to the metatarsus (the five bones of the foot between tarsals and phalanges).

Microfilament   Rodlike, protein filament about 6 nanometers (nm) in diameter. Contractile units in muscle cells; provides support, shape, and movement in nonmuscle cells.

Mineral   Inorganic, homogeneous solid substance that may perform a function vital to life. Remodeling of bone requires calcium, phosphorus, magnesium, and manganese.

Mineralization   *See* Calcification.

Mitochondrion   A double-membraned organelle where nearly all of the cell's ATP is produced. Abundant in muscle and liver cells.

Motor area   The region of the cerebral cortex (of the brain) that governs muscular movement; especially the precentral gyrus of the frontal lobe.

Motor end plate   Portion of the sarcolemma of a muscle fiber adjacent to an axon terminal of a neuron. Contains membrane receptors for neurotransmitter. Located where muscle action potentials are initiated.

Motor neuron   A neuron that conveys nerve impulses from the brain and spinal cord to effectors (muscle or glands). Also called an *efferent neuron*.

Motor unit   A motor neuron together with the muscle fibers it stimulates. The ratio of neuron to muscle fibers determines the degree of muscular control.

Multiunit smooth muscle tissue   One of two kinds of smooth muscle tissue. Consists of individual fibers, each with its own varicosity (like a synaptic end bulb) from which neurotransmitter is released. There are few gap junctions between neighboring fibers; stimulation of a fiber usually causes contraction of only that fiber.

Muscle   An organ composed of one of three types of muscle tissue : skeletal, cardiac, or smooth. Specialized for contraction to produce voluntary or involuntary movement of parts of the body.

Muscle action potential   A change in membrane potential in the sarcolemma of muscle cells. The combining of acetylcholine with receptors in the sarcolemma alters the membrane permeability to sodium ions, triggering an action potential. These action potentials self-propagate along the sarcolemma and along the transverse tubules to the sarcoplasmic reticulum, causing the release of calcium.

Muscle fiber   A muscle cell.

Muscle fatigue   Inability of a muscle to maintain its strength of contraction or tension; may be related to insufficient oxygen, depletion of glycogen, or lactic acid buildup. Also called *fatigue*.

Muscle spindle   An encapsulated receptor in a skeletal muscle, consisting of specialized muscle fibers and nerve endings. It is stimulated by changes in length or tension of muscle fibers. It provides information about body position or movement, so it is classified as a proprioceptor. Also called a *neuromuscular spindle*.

Muscle tissue   Tissue specialized for contraction. There are three types of muscle tissue : skeletal, smooth, and cardiac.

Muscle tone   A sustained, partial contraction of portions of a skeletal muscle in response to activation of stretch receptors.

Muscularis   A muscular layer (coat or tunic) of an organ.

Muscularis mucosae   A thin layer of smooth muscle fibers located in the outermost layer of the lining (mucosa) of the gastrointestinal tract.

Myasthenia gravis   Weakness and fatigue of skeletal muscles due to a decreased number of ACh receptors at the motor end plate.

Myofibril   Longitudinal bundle of thick and thin filaments arranged in sarcomeres (repeating patterns); located in the sarcoplasm of a skeletal muscle fiber.

Myoglobin   The oxygen-binding, iron-containing protein found in the sarcoplasm of muscle fibers; gives the red color to muscle. Stores oxygen for aerobic respiration.

Myogram   The record or tracing produced by the myograph, the apparatus that measures and records the effects of muscular contractions.

Myology   The study of the muscles.

Myometrium   The smooth muscle layer of the uterus.

Myoneural junction   *See* Neuromuscular junction.

Myosin   The contractile protein that makes up the thick filaments of muscle fibers.

Nasal septum   A vertical partition that separates the nasal cavity into right and left sides. It is composed of bone (the vomer and the perpendicular plate of the ethmoid) and cartilage, covered with a mucous membrane.

Neck   The part of the body connecting the head and the trunk. Constricted portion of a long bone near the head; an example is the neck of the femur.

Nerve impulse   A series of action potentials propagated along a nerve fiber (axon or dendrite).

Neuromuscular junction   The area of contact between the synaptic end bulbs of a motor neuron and a portion of the sarcolemma of a muscle fiber (motor end plate). Also called a *myoneural junction*.

Neurotransmitter   One of a variety of molecules synthesized within the axon terminal and released into the synaptic cleft in response to a nerve impulse. It interacts with receptors on the membrane of the adjacent cell, altering membrane permeability to ions. The adjacent cell may be another neuron, a muscle cell, or a gland cell. Also called a *transmitter substance*.

Nucleus pulposus   A soft, pulpy, highly elastic substance in the center of an intervertebral disc.

Orbit   A bony, pyramid-shaped cavity of the skull that holds an eyeball.

Orifice Any opening or aperture.

Origin The attachment of a muscle tendon to a stationary bone; the end of a muscle opposite the insertion.

Os coxae *See* Coxal bones.

Osseous Bony.

Ossicle *See* Auditory ossicle.

Ossification Formation of bone. Also called *osteogenesis*.

Osteoblast Cell formed from an osteoprogenitor cell that participates in bone formation by secreting some organic components and inorganic salts.

Osteoclast A large multinuclear cell that develops from a monocyte (type of white blood cell) and destroys or resorbs bone tissue.

Osteocyte A mature bone cell that maintains the daily activities of bone tissue.

Osteogenic layer The inner layer of the periosteum that contains cells responsible for forming new bone during growth and repair.

Osteology The study of bones.

Osteon The basic unit of structure in compact bones. Consists of a central canal that contains blood vessels and nerves, concentric rings of matrix (lamellae), small spaces (lacunae) containing bone cells (osteocytes), and small canals (canaliculi) that carry nutrients and wastes. Also called an *Haversian system*.

Osteoprogenitor cell Stem cell derived from mesenchyme that can divide by mitosis and differentiate into an osteoblast.

Oxygen debt *See* Recovery oxygen consumption.

Palate The roof of the mouth. The horizontal structure separating the oral and the nasal cavities.

Paranasal sinus A mucus-lined air cavity in a skull bone that communicates with the nasal cavity. Paranasal sinuses are located in the frontal, maxillary, ethmoid, and sphenoid bones.

Parathyroid hormone (PTH) A hormone secreted by the parathyroid glands that decreases blood phosphate level and increases blood calcium level.

Passive tension The tension generated by elastic components in a muscle. Elastic components include elastic filaments, connective tissue around the muscle fibers, and tendons that attach muscle to bone. Elastic components stretch slightly before they start to relay the force or tension being generated by the sliding filaments. Also called *external tension*.

Patellar Pertaining to the patella (kneecap).

Pectoral Pertaining to the chest or breast.

Pectoral girdle Each of the two pectoral girdles consists of two bones : scapula and clavicle.

Pelvic cavity Inferior portion of the abdominopelvic cavity. It contains the urinary bladder, sigmoid colon, rectum, and internal female and male reproductive structures.

Pelvic diaphragm The muscles of the pelvic floor together with the fascia covering their external and internal surfaces.

Pelvic floor *See* Perineum.

Pelvic girdle The pelvic girdle consists of the two hipbones (the coxal bones).

Pelvis Includes the pelvic girdle, sacrum, and coccyx.

Perforating canal A minute passageway through which nerves and blood vessels from the periosteum penetrate into compact bone. Also called *Volkmann's canal*.

Perichondrium The membrane that covers cartilage.

Perimysium Invagination of the epimysium that divides muscles into bundles of muscle fibers (fascicles).

Perineum The entire outlet of the pelvis; a diamond-shaped area at the lower end of the trunk between the thighs and buttocks. It is bordered anteriorly by the pubic symphysis, laterally by the ischial tuberosities, and posteriorly by the coccyx. In males : the region between the anus and scrotum. In females : the region between the anus and vulva. Also called the *pelvic floor*.

Periosteum The membrane that covers bone. It consists of connective tissue, osteoprogenitor cells, and osteoblasts. It is essential for bone growth, repair, and nutrition.

Phalanx (plural : phalanges) The bone of a finger or toe.

Phosphagen system Creatine phosphate and ATP constitute the phosphagen system and provide enough energy for muscles to contract maximally for about 15 seconds. Creatine phosphate can transfer its high-energy phosphate to ADP, forming ATP and creatine.

Pivot joint A synovial joint in which a rounded, pointed, or conical surface of one bone articulates with a ring formed partly by another bone and partly by a ligament. Examples include : the joint between the atlas (1st cervical vertebra) and the axis (2nd cervical vertebra); the joint between the proximal ends of the radius and ulna. Also called a *trochoid joint*.

Plantar flexion Bending the foot in the direction of the plantar surface (sole).

Pluripotent hematopoietic stem cell Immature stem cell in bone marrow that gives rise to precursors of all the different mature blood cells. Also called a *hemocytoblast*.

Posterior Nearer to the back; opposite of anterior. Also called *dorsal*.

Power stroke The action that causes thin filaments to slide past thick filaments during muscle contraction. Myosin cross-bridges swivel toward the center of the sarcomere, like the oars of a boat.

Primary curves (of vertebral column) The thoracic and sacral curves of the vertebral column.

Primary ossification center A region where bone tissue will replace most the cartilage. Osteoblasts deposit bone matrix over the remnants of calcified cartilage, forming spongy bone trabeculae.

Prime mover The muscle directly responsible for producing the desired movement. Also called the *agonist*.

Process A type of bone surface marking. Any prominent projection of bone. A vertebra has 7 vertebral processes.

Pronation A movement of the forearm in which the palm of the hand is turned posteriorly or inferiorly.

Proprioception The awareness of the precise position of body parts.

Proprioceptor A receptor located in muscles, tendons, or joints that provides information about body position and movements. Examples are muscle spindles and tendon organs.

Protraction The movement of the mandible or shoulder girdle forward on a plane parallel to the ground.

Proximal Nearer to the point of origin; away from the hand of the upper extremity or away from the foot of the lower extremity.

Pubic symphysis A slightly movable cartilaginous joint between the anterior surfaces of the hipbones. Also called *symphysis pubis.*

Quadriceps   Having four heads of origin. Four thigh muscles that extend the leg are called the quadriceps : rectus femoris, vastus intermedius, vastus lateralis, and vastus medialis.

Range of motion (ROM)   The maximum ability to move the bones of a joint through an arc. The closer a muscle attaches to a joint, the greater the ROM and speed of movement. The closer a muscle attaches to a joint, the less the mechanical advantage achieved by lever action. Strength and ROM vary inversely.

Recovery oxygen consumption   Extra oxygen taken into the body after exercise. This oxygen is used to convert lactic acid back to pyruvic acid, re-establish glycogen stores, resynthesize creatine phosphate and ATP, and replace oxygen removed from myoglobin. Also called *oxygen debt.*

Recruitment   The process of increasing the number of active motor units during muscle contraction; increases the strength of a contraction. Also called *motor unit summation.*

Red marrow   A type of marrow found in the spaces of spongy bone. Consists of blood cells in immature stages, adipose cells, and macrophages. It produces red blood cells, white blood cells, and platelets.

Red muscle fibers   Muscle fibers that contain large amounts of myoglobin, an oxygen-storing protein. They also contain more mitochondria and more capillaries than white muscle fibers.

Refractory period   A time during which an excitable cell (muscle or nerve cell) cannot respond to a stimulus that is usually adequate to evoke an action potential.

Relaxation period   The third phase of a twitch contraction. It is caused by the active transport of calcium ions back into the sarcoplasmic reticulum, which results in relaxation of the muscle fiber. Lasts 10 to 100 msec.

Relaxin   A female hormone produced by the ovaries that relaxes the pubic symphysis and helps dilate the uterine cervix to ease delivery of a baby.

Remodeling   Replacement of old bone by new bone tissue.

Resistance (R)   The force that opposes movement in a lever system.

Resorption   Bone loss due to the activity of osteoclasts.

Retraction   The movement of a protracted part of the body backward on a plane parallel to the ground. An example is pulling the lower jaw back in line with the upper jaw.

Rigor mortis   State of partial contraction of muscles following death. Lack of ATP causes cross-bridges of thick filaments to remain attached to thin filaments, preventing relaxation.

Rotation   Moving a bone around its own axis.

Sacral promontory   The superior surface of the first sacral vertebra that projects anteriorly into the pelvic cavity.

Sacrum   The 5 sacral vertebrae, which are fused.

Saddle joint   A synovial joint in which the articular surface of one bone is saddle shaped. An example is the joint between the trapezium (a carpal bone) and the metacarpal of the thumb. Also called a *sellaris joint.*

Sarcolemma   The cell membrane of a muscle fiber, especially a skeletal muscle fiber.

Sarcomere   A contractile unit in a striated muscle fiber. It extends from one Z disc (Z line) to the next Z disc. The distance between Z discs shortens when a muscle fiber contracts.

Sarcoplasm   The cytoplasm of a muscle fiber.

Sarcoplasmic reticulum   A network of saccules and tubules surrounding myofibrils of a skeletal muscle fiber; it is comparable to endoplasmic reticulum. When stimulated by an action potential it releases calcium ions that trigger the contraction of nearby sarcomeres.

Secondary curves (of vertebral column)   The cervical and lumbar curves of the vertebral column.

Secondary ossification center   Centers in developing bone that usually appear around the time of birth. Bone development is similar to that in the primary ossification centers.

Sellaris joint   *See* Saddle joint.

Sella turcica (Turk's saddle)   A depression on the superior surface of the sphenoid bone that houses the pituitary gland.

Sensory neuron   A neuron that conducts impulses from a sensory receptor into the central nervous system (brain or spinal cord). Also called an *afferent neuron.*

Sesamoid bones   Small bones usually found in tendons.

Shoulder   A ball-and-socket joint where the humerus joins the scapula.

Single-unit smooth muscle tissue   One of two kinds of smooth muscle tissue. The fibers form large networks because they contain gap junctions. When a fiber is stimulated, the muscle action potential spreads to neighboring fibers, which then contract as a single unit. Also called *visceral smooth muscle tissue.*

Skeletal muscle   An organ specialized for contraction, composed of striated muscle fibers (cells), supported by connective tissue. A skeletal muscle is attached to a bone by a tendon (fibrous band) or an aponeurosis (fibrous sheet composed of collagenous bundles).

Skeletal muscle tissue   Cylindrical, striated fibers with many peripheral nuclei; under voluntary control.

Skull   The skeleton of the head consisting of the cranial and facial bones.

Sliding-filament mechanism   The most commonly accepted explanation for muscle contraction in which actin and myosin filaments move into interdigitation with each other, decreasing the length of the sarcomeres.

Smooth muscle tissue   There are two kinds of smooth muscle tissue: single-unit (visceral) and multiunit. Single-unit smooth muscle tissue is found in the walls of small arteries, veins, and hollow organs (stomach, intestines, uterus, and urinary bladder). Multiunit smooth muscle tissue is found in the walls of large arteries, airways (bronchioles) to the lungs, arrector pili muscles of the skin, and radial and circular muscles of the iris that adjust pupil size.

Somatic   Pertaining to the body, especially the outer walls and framework of the body (skin, skeletal muscles, tendons, and joints). For example, nerve fibers of the somatic nervous system extend from the brain and spinal cord to skeletal muscles and skin.

Somite   Block of mesodermal cells in a developing embryo that is divided into three parts : a myotome, which forms most of the skeletal muscles; a dermatome, which forms connective tissues; and sclerotome, which forms the vertebrae.

Spheroid joint   *See* Ball-and-socket joint.

Sphincter   A circular muscle used for constricting an orifice.

Spinal nerve   One of the 31 pairs of nerves that originate on the spinal cord.

Spinous process   A sharp or thornlike projection of a bone. The sharp ridge running diagonally across the posterior surface of the scapula. Also called a *spine*.

Spongy bone   Consists of trabeculae surrounding many spaces filled with red marrow. It forms most of the structure of short, flat, and irregular bones, and the epiphyses (ends) of long bones. The term *cancellous* denotes the honeycomb structure of spongy bone.

Staircase effect   *See* Treppe.

Sternum   Long, flat bone forming the middle part of the anterior wall of the rib cage. It articulates with the clavicles and with the costal cartilages of the first seven pairs of ribs.

Subcutaneous layer   *See* Superficial fascia.

Summation   The increased strength of muscle contraction that results when stimuli follow in rapid succession.

Superficial   Located on or near the surface of the body.

Superficial fascia   A continuous sheet of fibrous connective tissue between the dermis of the skin and the deep fascia of the muscles. Also called *subcutaneous layer*.

Superficial inguinal ring   A triangular opening in the aponeurosis (fibrous sheet) of the external oblique muscle; represents the termination of the inguinal canal.

Superior   Toward the head or upper part of a structure. Also called *cephalad* or *craniad*.

Supination   A movement of the forearm in which the palm of the hand is turned upward.

Sutural bone   A small bone located within a suture between certain cranial bones. Also called *Wormian bone*.

Suture   An immovable fibrous joint in the skull where bone surfaces are closely united.

Symphysis   A line of union. A slightly movable cartilaginous joint such as the pubic symphysis between the anterior surfaces of the hipbones.

Synapse   The junction between two neurons.

Synaptic cleft   The narrow gap that separates the axon terminal of one nerve cell from another nerve cell or muscle fiber. Neurotransmitter diffuses across the synaptic cleft to affect the postsynaptic cell.

Synaptic end bulb   The expanded end of an axon terminal. It contains synaptic vesicles filled with neurotransmitter.

Synaptic vesicle   Membrane-enclosed sac in a synaptic end bulb that stores neurotransmitter.

Synarthrosis   An immovable joint. Includes: suture (between cranial bones), gomphosis (between roots of teeth & jawbone), and synchondrosis (between true ribs & sternum).

Synchondrosis   A cartilaginous joint in which the connecting material is hyaline cartilage; example is the joint between true ribs and the sternum.

Syndesmosis   A joint in which the articulating bones are united by dense fibrous tissue; an example is the distal ends of the tibia and fibula. Classified as a slightly movable joint (amphiarthrosis).

Synergist   A muscle that assists the prime mover by reducing undesired action or unnecessary movement; stabilizes the action of the prime mover.

Synostosis   A joint in which the dense fibrous connective tissue that unites bones at a suture has been replaced by bone, resulting in a complete fusion across the suture line.

Synovial cavity   The space between the articulating bones of a synovial (diarthrotic) joint, filled with synovial fluid. Also called a *joint cavity*.

Synovial fluid   Secretion of synovial membranes that lubricates joints and nourishes articular cartilage.

Synovial joint   A fully movable or diarthrotic joint in which a synovial (joint) cavity is present between the two articulating bones.

Synovial membrane   The inner of the two layers of the articular capsule of a synovial joint; it is composed of areolar connective tissue that secretes synovial fluid into the synovial (joint) cavity.

Tarsal   Pertaining to the tarsus (the seven small bones between the leg bones and the metatarsus).

Tendon   A white fibrous cord of dense, regularly arranged connective tissue that attaches muscle to bone.

Tendon organ   A proprioceptive receptor sensitive to changes in muscle tension and force of contraction. Found chiefly near the junction of tendons and muscles. Also called a *Golgi tendon organ*.

Tetanus   A smooth, sustained contraction produced by a series of very rapid stimuli to a muscle. Also the name of an infectious disease characterized by tonic muscle spasms, exaggerated reflexes, lockjaw, and arching of the back.

Thigh   The portion of the lower extremity between the hip and the knee.

Thoracic cage   The ribs and sternum. Also called *rib cage*.

Thoracic cavity   The component of the ventral cavity that is superior to the diaphragm. It contains the lungs and the mediastinum.

Thoracic vertebrae   Twelve vertebrae that articulate with the twelve ribs.

Thorax   The chest.

Trabeculae (singular : trabecula)   Lamellae arranged in an irregular latticework of thin plates; the structure of spongy bone. The spaces between the trabeculae are filled with red bone marrow.

Treppe   The gradual increase in the amount of contraction by a muscle caused by rapid, repeated stimuli of the same strength. Also called *staircase effect*.

Triad   A complex of three units in a muscle fiber. Composed of a transverse tubule and the segments of sarcoplasmic reticulum on both sides of it.

Trochanter   A large projection found only on the proximal end of the femur.

Trochoid joint   *See* Pivot joint.

True ribs   The first 7 pairs of ribs. They have a direct individual attachment to the sternum by a strip of hyaline cartilage called costal cartilage.

Trunk   The part of the body to which the upper and lower extremities are attached. Also called the *torso*.

Tubercle   A small, rounded process. An example is the greater tubercle of the humerus.

Tuberosity   A large, rounded, usually roughened process. An example is the tibial tuberosity of the tibia.

Twitch contraction   A brief contraction of all the muscle fibers in a motor unit in response to a single action potential in its motor neuron.

Upper extremity   The appendage attached at the shoulder girdle, consisting of the arm, forearm, wrist, hand, and fingers.

Urogenital diaphragm   The part of the urogenital triangle consisting of the deep transverse perineus muscle, the urethral sphincter, and a fibrous membrane.

Urogenital triangle   The diamond-shaped perineum may be divided into anterior and posterior triangles by a line drawn between the ischial tuberosities. The anterior triangle is called the urogenital triangle and contains the external genitals.

Varicosity   A swollen portion of an axon branch. Varicosities are found on the axon branches of neurons that stimulate smooth muscle cells.

Ventral   Nearer to the front; opposite of dorsal. Also called *anterior*.

Vertebra (plural : vertebrae)   One of the 33 bones that make up the vertebral column.

Vertebral arch   The posterior portion of a vertebra. It forms an arch that attaches to the vertebral body. It contains 7 processes, 2 pedicles, and 2 laminae.

Vertebral body   The heavy, anterior portion of a vertebra that supports most of the weight of the vertebral column. The vertebral bodies of adjoining vertebrae are separated by pads of cartilage called intervertebral discs.

Vertebral canal   A cavity within the vertebral column formed by the vertebral foramina of all the vertebrae and containing the spinal cord. Also called the *spinal canal*.

Vertebral column   The 33 vertebrae; there are just 26 separate bones, since the 5 sacral vertebrae are fused and the 4 coccygeal vertebrae are fused. Encloses and protects the spinal cord and serves as a point of attachment for the ribs and back muscles. Also called the *spine*, *spinal column*, or *backbone*.

Vertebral foramen   The hole inside each vertebra where the spinal cord passes through. The vertebral foramina of all the vertebrae form a cavity called the vertebral canal, which contains the spinal cord.

Vertebral processes   Portions of the bone that extend out from the vertebral arch. The spinous and transverse processes serve as points of attachment for muscles; the articular processes have surfaces called facets that form joints with other vertebrae and ribs.

Visceral   Pertaining to the organs or to the coverings of the organs of the ventral body cavity.

Visceral muscle   An organ specialized for contraction, composed of smooth muscle fibers. Located in the walls of the hollow internal structures.

Visceral smooth muscle tissue   One of two kinds of smooth muscle tissue. The fibers form large networks because they contain gap junctions. When a fiber is stimulated, the muscle action potential spreads to neighboring fibers, which then contract as a single unit. Also called *single-unit smooth muscle tissue*.

Vitamins   Organic molecules which must be present in trace amounts to maintain normal health. They must be part of the diet, since they cannot be synthesized in the body. Bone remodeling requires vitamins A , and B-12, C, and D.

Volkmann's canal   *See* Perforating canal.

Wave summation   The increased strength of muscle contraction that results when stimuli follow in rapid succession. Also called *temporal summation*.

White muscle fibers   Fibers that have a low content of

myoglobin, an oxygen-storing protein.

Wormian bone   *See* Sutural bone.

Xiphoid process   The lowest portion of the sternum.

Yellow marrow   Marrow found in the medullary cavity of long bones; consists primarily of adipose (fat) cells and a few scattered blood cells.

Zone of calcified cartilage   Zone in the epiphyseal plate that consists mostly of dead cells because the matrix around them has calcified.

Zone of hypertrophic cartilage   Zone in the epiphyseal plate that consists of very large chondrocytes arranged in columns. Also called *maturing cartilage*.

Zone of proliferating cartilage   Zone in the epiphyseal plate that consists of chondrocytes that divide to replace those that die at the diaphyseal surface of the epiphyseal plate.

Zone of resting cartilage   Zone in the epiphyseal plate that consists of small, scattered chondrocytes. The cells anchor the epiphyseal plate to the bone of the epiphysis.

Z discs   Narrow, plate-shaped regions of dense material that separate adjacent sarcomeres. Thin filaments (actin) extend toward the center of each sarcomere from anchoring points within the Z discs. Also called *Z lines*.

Bibliography

Dorland, William Alexander.   *Dorland's Illustrated Medical Dictionary,* 27th ed.
Philadelphia : W. B. Saunders, 1988.

Fowler, Ira.   *Human Anatomy.*
Belmont, California : Wadsworth, 1984.

Ganong, William F.   *Review of Medical Physiology*, 15th ed.
Norwalk, Connecticut : Appleton & Lange, 1991.

Goldberg, Stephen.   *Clinical Anatomy Made Ridiculously Simple.*
Miami, Florida : MedMaster, 1984.

Junqueira, L. Carlos, Jose Carneiro, and Robert O. Kelley.   *Basic Histology*, 6th ed.
Norwalk, Connecticut : Appleton & Lange, 1989.

Kapit, Wynn and Lawrence M. Elson.   *The Anatomy Coloring Book.*
New York : Harper & Row, 1977.

Melloni, B. J., Ida Dox, and Gilbert Eisner.   *Melloni's Illustrated Medical Dictionary,* 2nd ed.
Baltimore : Williams & Wilkins, 1985.

Moore, Keith L.   *Clinically Oriented Anatomy,* 3rd ed.
Baltimore : Williams & Wilkins, 1992.

Netter, Frank H.   *Atlas of Human Anatomy.*
Summit, New Jersey : Ciba-Geigy, 1989.

Tortora, Gerard J. and Sandra Reynolds Grabowski. *Principles of Anatomy and Physiology,* 7th ed.
New York : HarperCollins, 1993.

Vander, Arthur J., James H. Sherman, and Dorothy S. Luciano.   *Human Physiology,* 5th ed.
New York : McGraw-Hill, 1990.